The Origin of Time

SUNY series in Contemporary Continental Philosophy

Dennis J. Schmidt, editor

The Origin of Time

Heidegger and Bergson

Heath Massey

SUNY
PRESS

Published by State University of New York Press, Albany

Printed in the United States of America

For information, contact State University of New York Press, Albany, NY
www.sunypress.edu

Production, Diane Ganeles
Marketing, Kate R. Seburyamo

Library of Congress Cataloging-in-Publication Data

Massey, Heath, author.
The origin of time : Heidegger and Bergson / Heath Massey.
pages cm. — (SUNY series in contemporary continental philosophy)
Includes bibliographical references and index.
ISBN 978-1-4384-5531-0 (hardcover : alk. paper)
ISBN 978-1-4384-5532-7 (pbk. : alk. paper)
ISBN 978-1-4384-5533-4 (ebook)
1. Time. 2. Heidegger, Martin, 1889–1976. 3. Bergson, Henri, 1859–1941.
4. Phenomenology. I. Title.

BD638.M364 2015
115—dc23 2014013136

10 9 8 7 6 5 4 3 2 1

To my parents.

Contents

Acknowledgments

My heartfelt thanks to my colleagues in the Department of Philosophy and Religious Studies at Beloit College: Phil Shields, Matt Tedesco, Robin Zebrowski, Debra Majeed, and Natalie Gummer. I am grateful to those who generously gave their time to read parts or all of the manuscript: Karin Fry, Topper Scott, Len Lawlor, David Morris, Alia Al-Saji, and Michael Kelly. Thanks also to Scott Lyngaas and Nataša Bašić for their feedback on an earlier draft of Chapter 2. I owe a great debt to all my professors at the University of Memphis, and in particular to Tina Chanter, Tom Nenon, Robert Bernasconi, and especially Len Lawlor for guidance on this project in its early stages. I benefited tremendously from the community of graduate students at Memphis, including Amit Sen, Chris Fox, Josh Glasgow, Valentine Moulard-Leonard, Rob Wade, and my dear friend Steve Tammelleo. I appreciate the participants of the 2003 *Collegium Phaenomenologicum* for helping to shape my thinking about Bergson and his relationship to phenomenology. Special thanks to John Sallis for sharing a photocopy of Heidegger's personal copy of Bergson's *Essai*. Many thanks to Dennis Schmidt for supporting this project; to two anonymous reviewers for providing valuable feedback on the manuscript; and to Andrew Kenyon, Diane Ganeles, Laura Tendler, and Kate Seburyamo at SUNY Press for turning it into a book. For assistance and support at the final stages, my thanks to Becky Moffett, Barb Uebelacker, and Ann Davies. I could not have finished this project without the generosity of Kathryn Boone and Elva Tyson. For their friendship and moral support, I deeply appreciate Charles and Christa Westerberg and John and Ellen Morgan. Many thanks to Ted Ammon, Michael Mitias, and Steve Smith for their mentorship and inspiration, and to David Hiley for his guidance. I am more grateful than I can say to my parents, David and Karen, and my brother, Seth, for their constant encouragement and support. Finally, words cannot express the depth of my gratitude to my son, Theo, and my wife, Elizabeth, for all their love, patience, and understanding.

A version of Chapter 1 appeared as "On the Verge of *Being and Time*: Before Heidegger's Dismissal of Bergson," *Philosophy Today* 54 (2010): 138–152.

Abbreviations

BPPa	*Basic Problems of Phenomenology* (GA 58), Martin Heidegger
BPPb	*The Basic Problems of Phenomenology* (GA 24), Martin Heidegger
BT	*Being and Time* (GA 2), Martin Heidegger
CT	*The Concept of Time* (GA 64), Martin Heidegger
CE	*Creative Evolution*, Henri Bergson
CM	*The Creative Mind*, Henri Bergson
HCT	*History of the Concept of Time* (GA 20), Martin Heidegger
LQT	*Logic: The Question of Truth* (GA 21), Martin Heidegger
MM	*Matter and Memory*, Henri Bergson
MFL	*The Metaphysical Foundations of Logic* (GA 26), Martin Heidegger
P	*Pathmarks* (GA 9), Martin Heidegger
PIE	*Phenomenology of Intuition and Expression* (GA 59), Martin Heidegger
S	*Supplements* (GA 1, GA 80), Martin Heidegger
TDP	*Toward the Definition of Philosophy* (GA 56/57), Martin Heidegger
TFW	*Time and Free Will*, Henri Bergson

In all citations, the page number of the English translation is followed by that of the original French or German.

Introduction

Thinking Through Time

The figure of Martin Heidegger looms large over philosophy today, and a tremendous amount of attention is devoted to his work. The same is not true of Henri Bergson, whose fortunes have been mixed since his philosophy flourished for much of the first half of the twentieth century and then fell so far out of fashion as to be virtually forgotten. Given the amount of influence Bergson had on philosophy and culture during his lifetime, it is surprising how far and how fast his star fell. In France, at least, many of the thinkers who were in the best position to keep Bergson at the center of philosophical discussion were more interested in Heidegger and his teacher Edmund Husserl. Instead of Bergsonism, the philosophies that grew out of the soil that Bergson prepared called themselves existential phenomenology, deconstruction, or philosophy of difference. Bergsonism became associated with the philosophical establishment and old-fashioned metaphysics. Duration, *élan vital*, and intuition—unless preceded by "Husserlian"—became philosophical four-letter words.

Recently, though, there has been a significant renewal of interest in Bergsonism and a greater recognition of his influence on twentieth-century philosophy.[1] Much of the contemporary interest in Bergson is due to his influence on Gilles Deleuze, but there has also been an increase in scholarly attention devoted to Bergson's relationship to phenomenology. The question of Heidegger's debt to Bergson was raised long ago, but it has remained largely unanswered.

In *Cinema I*, Deleuze writes, "The only resemblance between Bergson and Heidegger—and it is a considerable one—lies here: both base the specificity of time on a conception of the open."[2] Without attributing a direct influence, Maurice Merleau-Ponty nevertheless described Bergsonian duration as having a " 'singular nature' . . . which makes it at once my manner of being and a universal dimension for other beings."[3] Emmanuel Levinas, when asked once what he thought was Bergson's principal contribution to philosophy, responded:

> The theory of duration. The destruction of the primacy of clock time; the idea that the time of physics is merely derived. Without this affirmation of the somehow "ontological" and not merely psychological priority of the duration irreducible to linear and homogeneous time, Heidegger would not have been able to venture his conception of Dasein's finite temporalization, despite the radical difference which separates, of course, the Bergsonian conception of time from the Heideggerian conception. The credit goes back to Bergson for having liberated philosophy from the prestigious model of scientific time.[4]

Jacques Derrida also suggests, in light of the various references to Bergson in Heidegger's early lecture courses, that "we know now that Heidegger read him more than his texts would lead one to think."[5] While these French thinkers recognized Bergson's influence decades ago, the publication of Heidegger's *Gesamtausgabe* has made it possible to develop a clearer picture of both the resemblances—which, contrary to Deleuze's provocative claim, are many—and the genuine differences in their thinking.

The relationship between Heidegger's thought and Bergson's is a difficult question for several reasons. One is that Heidegger evokes Bergson early in *Being and Time* in association with Aristotle and the "traditional concept of time," only to dismiss him hastily in a footnote near the end. Heidegger's brief discussion of Bergson is geared toward explaining why Bergson fails in his attempts to think more radically about time. Heidegger concludes with a promise to come back to Bergson's interpretation of time in the second part of *Being and Time*, but because he never published part two, this is a promise he was not able to keep. Another reason for the difficulty of this question is that it is not clear where Bergson's influence on Heidegger begins and that of a number of other, better-known philosophers begins. Husserl, Dilthey, Kant, and Aristotle, for instance, were all subjects of major studies by Heidegger during the period in which he wrote *Being and Time.* Heidegger developed his own interpretation of time with regard to these thinkers and others, including Kierkegaard, St. Paul, and Augustine. Although Heidegger mentions Bergson as a contemporary heir to Aristotle with respect to time on several occasions early in *Being and Time*, he concludes division two with a detailed critique of Hegel's interpretation of time instead. Any influence Bergson exerted on Heidegger's thinking was combined with that of a host of others, most of whom Heidegger was more willing to engage with openly. Without a decisive confrontation with Bergson, and with many other influences in the mix, the task of determining Heidegger's debt becomes a formidable one.

While I would not claim that Heidegger's thinking about time was influenced solely or even primarily by Bergson, I do hope to show that the casual, offhand way that Heidegger deals with Bergson in *Being and Time* conceals a deep, almost subterranean influence. Even though Heidegger appears to be most concerned with distancing himself from Bergson, his strategy for dealing with time displays a profound engagement with Bergson. Over a period of slightly more than a decade prior to *Being and Time*, Heidegger develops his thinking about temporality with Bergson as a continual point of reference. Early on, Heidegger follows Bergson in the way he thinks about time by appropriating several crucial insights of Bergson's. In the years just prior to the publication of *Being and Time*, he places Bergson on a level with Kant and Aristotle for his attempt to transform our understanding of time, but he also criticizes Bergson for failing to overcome the traditional concept of time. However, like Bergson, Heidegger seeks to show that certain assumptions about time are at the basis of some of philosophy's most fundamental and persistent problems.[6] As such, Heidegger and Bergson are united in not only challenging the way philosophers have traditionally thought *about* time, but also indicating new directions for philosophy by thinking *through* time.

Bergson was born in 1859, the same year as Husserl. His first book, *Essai sur les données immédiates de la conscience* (translated as *Time and Free Will*), was published in 1889, the year that Heidegger was born. Bergson's attempt in that essay to rethink time in terms of duration preceded not only Heidegger's interpretation of temporality in *Being and Time*, published in 1927, but also Husserl's famous lectures on the phenomenology of time-consciousness from 1904–5.[7] Bergson's philosophy was already being taught in Germany by the time Heidegger was a doctoral student, and Heidegger studied Bergson with Heinrich Rickert at the University of Freiburg.[8] By the time Heidegger received his *Habilitation* degree in 1915, Bergson had published two other major works, *Matière et mémoire* (*Matter and Memory*, 1896) and *L'Évolution créatrice* (*Creative Evolution*, 1907), as well as numerous essays and lectures, including *Introduction à la métaphysique* ("Introduction to Metaphysics," 1903) and *L'Intuition philosophique* ("Philosophical Intuition," 1911). Bergson's lectures at the Collège de France had won him a huge popular following; he had been elected to the Académie française and his books had been placed on the Catholic *Index Librorum Prohibitorum* (both in 1914); and by the time *Being and Time* appeared, he had been awarded the Nobel Prize for Literature.[9]

From the beginning of his career, time was an especially crucial issue for Bergson and a major point of contention for him with the entire tradition of Western philosophy. While maintaining that time, change, growth, and novelty are fundamental to human experience and reality as such, Bergson noted that philosophers have usually sought to transcend the temporal. According to him: "Metaphysics dates from the day when Zeno of Elea pointed out the inherent contradictions in movement and change, as our intellect represents them," and to resolve these paradoxes, "metaphysics was led to seek the reality of things above time, beyond what moves and changes, and consequently outside what our senses and consciousness perceive" (CM 17/8). From Plato's divided line, to Spinoza's view *sub specie aeternitatis*, to McTaggert's A-series and B-series, philosophers have tended to privilege that which is timeless and to regard time as an obstacle, an appearance, or an illusion. But for Bergson, the desire to adopt a standpoint outside of time is responsible for many of the apparently insoluble problems of metaphysics.

If the aim of most philosophers has been to elevate the mind to a position from which the eternal can be contemplated, free from the illusions of temporal existence, Bergson calls for philosophy to resist this temptation. Thinking, for him, "is not a question of getting outside of time," but of getting "back into duration" (CM 31/1272). Contrary to conventional wisdom, it is not the temporal realm that is illusory, but rather our spatial representations of phenomena that are fundamentally temporal. In Bergson's view, we cannot grasp the unceasing flux of reality with static, ready-made concepts. Most philosophical problems grow out of certain intellectual habits acquired in response to vital needs and interests. Because these habits have shaped our thinking, he argues, it will take a concerted effort to break them. In a crucial passage from *Matter and Memory*, Bergson issues this challenge: "*Questions relating to subject and object, to their distinction and their union, should be put in terms of time rather than of space*" (MM 71/218). This is the clearest, most direct formulation of his strategy of attempting to resolve seemingly insoluble philosophical problems by breaking the habit of representing things spatially that should be understood temporally.

Heidegger similarly opposes the notion that time is an obstacle for philosophers to overcome in pursuit of eternal truths. On his reading of the history of philosophy, the ancient Greeks defined being with respect to time in a way that has been decisive for Western thought. In a lecture from the period during which he was writing *Being and Time*, he declares, "being in general, as the being of every being, must be conceived in terms of time. Or at any rate, according to the state of our present philosophical possibilities, being can be singularly understood in terms of time" (LQT 222/267). In the prologue to *Being and Time*, he begins his pursuit of the meaning of being on the basis of "the interpretation of *time* as the possible horizon for any understanding whatsoever of being" (BT xxix/1). For Heidegger, the fact that being is traditionally held to be timeless, and that philosophers typically distinguish between the temporal and the eternal as separate realms of being, raises the question of how and why time came to be viewed as something in which everything comes into being and passes away. The way that being is defined or interpreted in terms of time is problematic for many reasons, not the least of which is that it determines our way of thinking about the kind of being that we are. Consequently, "everything comes down to seeing the phenomenon of time itself in an original way" (LQT 173/205). This means that we need not a new concept of time, according to Heidegger, but rather a new understanding of the origin of time.

What right do philosophers have to speak of the "origin of time"? This sounds like a topic for cosmology, which would look to the beginning of the universe for answers about time's origin. Of course, the idea of the beginning of the universe involves a puzzle articulated by Augustine: if everything in the universe can be traced back to a starting point or primordial event from which all history has unfolded, then what came before that event? On a different interpretation, however, the "origin of time" could refer not to a primal event in the history of the universe, but to something about the human mind and the way it is constructed. Here, psychology would appear to be in the best position to investigate time's origin. Philosophical investigations of time have pursued both cosmological and psychological interpretations of its origin, depending on whether time is regarded as a feature of the world or the mind. In the introduction to his lectures on the phenomenology of time-consciousness, Husserl proposes yet another interpretation:

> The question about the essence of time thus leads back to the question about the *"origin" of time*. But this *question of origin* is directed toward the *primitive* formations of time-consciousness, in which the primitive differences of the temporal become constituted intuitively and properly as the original sources of all the evidences relating to time. The question of origin should not be confused with *the question about psychological origin*. . . . The question about empirical genesis is a matter of indifference as far as we are concerned; what does interest us are experiences with

> respect to their objective sense and descriptive content. . . . That these experiences are themselves fixed in objective time, *that they belong in the world of physical things and psychic subjects*, and that they have their place, their *efficacy*, their empirical being, and their origin in this world does not concern us and we know nothing about it.[10]

According to this way of thinking, the origin of time does lie in human experience, as psychologists who study it suppose, but it is not to be found in empirical facts and theories about the causes of that experience—that is, not in psychology. As Husserl presents it, phenomenology is concerned with essences rather than facts, and with rigorous descriptions of the structures and activities of consciousness rather than causal explanations. More precisely, it is concerned with *a priori* structures that are discovered through a close analysis of "temporal objects" and "time-constituting acts." For Husserl, the question is how our subjective experience of time, as well as the appearance of time as objective, arises from certain *a priori* structures of consciousness.

Bergson anticipates Husserl's departure from psychology to some extent by investigating time with regard to the "immediate data of consciousness." Heidegger departs even further from psychology by using the phenomenological method for an inquiry into being, avoiding as much as possible all talk of "consciousness" and "lived experience." Although Bergson did not belong to the phenomenological movement, instead developing his own method of "philosophical intuition," and Heidegger eventually ceased to identify himself as a phenomenologist, they both approached time, in the same spirit as Husserl, as not just one feature of our experience among many, but a particularly significant, fundamental, and *primordial* one.

In *Being and Time* and several of the lecture courses surrounding it, Heidegger criticizes Bergson for having tried but failed to grasp the phenomenon of *originary temporality*. According to Heidegger, Bergson's concept of duration represents an attempt to go beyond the traditional concept of time, going all the way back to Aristotle, to a more primordial time from which that concept is derived. In this respect, Heidegger's interpretation of temporality in *Being and Time* is an answer to Bergson, whom he criticizes for relying too heavily on Aristotle's definition of time, for neglecting the question of being, and for overstating the case against the ordinary concept of time. Heidegger links Bergson not only to Aristotle, but also to more recent thinkers such as Wilhelm Dilthey, Max Scheler, and Edmund Husserl, all of whom he criticizes for neglecting ontology. In a number of lectures and essays prior to *Being and Time*, Heidegger also mentions Bergson in connection with *Lebensphilosophie* (philosophy of life), expressing reservations about both his concept of life and his method. In the end, these attempts on Heidegger's part to distance himself from Bergson make it possible to see their proximity.

Both Bergson and Heidegger dealt with the issue of time, explicitly or implicitly, throughout their entire careers. However, Heidegger only directly engaged with Bergson during his early Freiburg and Marburg periods, from 1915 to 1928, so I have chosen to focus on works produced in those years. When Heidegger returned to Freiburg, he apparently abandoned any plans he might have had for a decisive confrontation with Bergson's thought. The decision to limit attention to Heidegger's early works has allowed me to focus tightly on his engagement with Bergson, but it has also resulted in some regrettable omissions. A more extensive treatment of temporality in Heidegger would have to consider *The Fundamental Concepts of Metaphysics*, his *Nietzsche* lectures, the *Beiträge*, the "Anaximander Fragment," and *Time and Being*, to name just a few. Focusing on Heidegger's early essays and lectures, *Being and Time* division two, and *The Basic Problems of Phenomenology* has made it possible for me to go into greater detail about Heidegger's struggle with and against Bergson. To clarify Bergson's role in *Being and Time*, I have offered a close reading of the sections of division two that deal most directly with time and temporality. Readers who are already well versed in Heidegger's account of the temporality of Dasein may want to skip those sections to get more quickly to my argument about what Heidegger appropriates from Bergson.

Although time is central to Bergson's thinking throughout his career, I have not attempted to provide a comprehensive study of duration in his works. Such a study would have to deal in greater depth with his most celebrated book, *Creative Evolution*; his confrontation with Einstein, *Duration and Simultaneity*; and his final and most neglected book, *The Two Sources of Morality and Religion*. Instead, I have offered close readings of Bergson's first two major works, *Time and Free Will* and *Matter and Memory*, both published prior to 1900. Heidegger's interpretation of Bergson deals almost exclusively with the former, where Bergson first develops the concept of duration, but I believe that much of what Heidegger found valuable about Bergson's thought appears in the latter. As these works are not very widely read today, I have been systematic in laying out Bergson's arguments, paying special attention to his strategy of posing questions in terms of time and rejecting assumptions that arise from thinking spatially about temporal phenomena. This has enabled me to respond to Heidegger's critique on the basis of a more nuanced interpretation of Bergson's thought and to show that his understanding of duration undergoes important changes as he develops his theory of memory. Readers already familiar with these works may want to skip some of the sections in Chapters 2 and 5 that are devoted to explicating Bergson's philosophy.

Chapter 1 focuses on Heidegger's critical engagement with Bergson in his lectures and essays prior to *Being and Time*. Surveying Heidegger's early works on time, from his 1915 *Habilitation* lecture to the 1925–6 lecture courses just prior to the publication of *Being and Time*, I show that Heidegger follows Bergson's footsteps by distinguishing between quantitative, homogeneous time and qualitative, heterogeneous time. Although he voices concerns about Bergson's method and his concept of life, Heidegger praises Bergson for recognizing the need to think more radically about time. Thus, Heidegger displays a deep ambivalence in his treatment of Bergson, crediting him with transforming the concept of time, but also attacking him for his dependence on the Aristotelian categories of quality and quantity, his uncritical acceptance of the notion of time as succession, and his misunderstanding of time as a representation of space. This survey of Heidegger's treatment of Bergson in his early works provides the background necessary to understand Bergson's role in *Being and Time*.

In Chapter 2, I evaluate Heidegger's critique of Bergson through a close reading of *Time and Free Will*, the main work that Heidegger cites as evidence of Bergson's dependence on Aristotle. With attention to the philosophical context and motives for Bergson's distinction between pure duration and time as a homogeneous medium, I show that Heidegger exaggerates the connection between Bergson's thinking and Aristotle's. In *Being and Time*, Heidegger condenses Bergson's philosophy of time into two formulas: "time is space" and duration is "qualitative succession." I argue that while there is some truth in these formulas, they oversimplify and distort Bergson's thinking. I demonstrate how Bergson's distinction between two kinds of multiplicity and two aspects of the self makes it possible for him to show that the problem of freedom arises from the assumption that time can be adequately represented in terms of space. Finally, I argue that Bergson's efforts to rethink time and the self in terms of duration anticipate Heidegger's own interpretation of temporality.

The subject of Chapter 3 is Bergson's role in *Being and Time*. Focusing first on Heidegger's explicit remarks about Bergson, I show that Heidegger's dismissal of Bergson rests on three major complaints: his reliance on Aristotelian categories, his neglect of ontology, and his treatment of the concept of time as illegitimate. Next, I focus on Heidegger's efforts to think more radically about time, which lead him to characterize originary temporality as a double movement reflected in the structure of care. Even though Heidegger avoids the term "duration" and Bergson's ways of characterizing it, he follows Bergson in seeking to uncover primordial time. When he does briefly discuss duration, Heidegger identifies it with one of the characteristics of "world time." With regard to his account of how the concept of time is derived from temporality, I show that the crux of Heidegger's critique is this: Bergson thinks that time is misunderstood because it is confused with space, but the issue is really the privilege of presence. However, while Heidegger casts Bergson as a foil for his own thinking, I argue that, like Bergson, Heidegger is concerned with disentangling temporality from spatiality and thinking about selfhood in temporal terms. Thus, Bergson plays a central role in Heidegger's interpretation of temporality.

In Chapter 4, I show that Heidegger continues to engage with Bergson in *The Basic Problems of Phenomenology*, again by subordinating his thinking about time to Aristotle's. First, I consider Heidegger's comments about Bergson and the traditional concept of time at the beginning of his discussion of temporality, where he reiterates his reasons both for taking Bergson seriously and for dismissing him, again with reference to Aristotle's concept of time. Then I evaluate Heidegger's critique of Bergson for misinterpreting Aristotle as reducing time to space. I show how Heidegger reads Aristotle not only as formulating the traditional concept of time, which covers up originary temporality, but also as catching a glimpse of the structure of temporality. Finally, I argue that Heidegger offers a selective reading of Aristotle to interpret him as having already anticipated Bergson's worry about confusing time with space. With this interpretation, Heidegger both confines Bergson's thought to the tradition inaugurated by Aristotle and uses Bergson's critique of that tradition as leverage to lift Aristotle above standard interpretations. The result is a strangely inverted history of the concept of time, in which Bergson appears as being thoroughly traditional and Aristotle appears as providing insights into temporality more radical than those of Heidegger's own contemporaries.

In Chapter 5, I examine Bergson's *Matter and Memory* and argue that, contrary to Heidegger's claims, Bergson actually does call the being of consciousness and the privilege of presence into question. First, I show how his theory of pure perception overturns the notion of consciousness as a subjective, inner realm of representation. Then I discuss how Bergson's theory of pure memory affirms the existence of unconscious memories, which are never present, and the integral survival of the past. I argue that Bergson not only rejects the equation of being with being present, but he also questions the temporal and ontological conditions of presence. This leads him to rethink the relationship between consciousness and duration and to argue that the duration of human experience is open to other rhythms of duration. I argue that Bergson's turn toward ontology transforms his understanding of time, and that Heidegger had to ignore this to portray him as concerned mainly with the consciousness of time.

1

Following Bergson's Footsteps

Time in Heidegger's Early Works

1. The Question of Time

Many readers of *Being and Time* have noticed that when Heidegger distinguishes between temporality and what philosophers call "time," he evokes Bergson's name along with Aristotle's. Yet the importance of Bergson's role in *Being and Time* is rarely acknowledged. It is tempting to conclude from the way Heidegger dismisses Bergson in a footnote near the end of division two that he was only interested in preventing his readers from confusing his view of time with Bergson's. The evidence from Heidegger's early works suggests otherwise. A survey of his treatments of time prior to the publication of *Being and Time* in 1927 reveals a sustained engagement with Bergson's thinking. With reference to the lecture courses and essays from this period, I aim to show, first, that not only did Heidegger consider Bergson a pivotal thinker with respect to time, but he also followed Bergson's footsteps in thinking about time in several crucial ways; and second, that his treatment of Bergson displays an ambivalence stemming from concerns about Bergson's understanding of life. Because Heidegger never lectured or wrote at length about Bergson, my case depends on evidence gathered from a variety of sources, some in which Bergson is mentioned by name and others in which he is not. By surveying the path leading to Heidegger's dismissal of Bergson, I hope to shed more light on Bergson's appearances in *Being and Time* and the role he plays in that work and in the early development of Heidegger's thinking about temporality.

Heidegger's initial remarks in *Being and Time* about Bergson are brief but provocative. In the prologue, Heidegger sets his sights on "the interpretation of *time* [*Zeit*] as the possible horizon for any understanding whatsoever of being" (BT xxix/1). He fleshes this out somewhat in the introduction, announcing, "The meaning of being of that being we call Dasein will prove to be *temporality* [*Zeitlichkeit*]," and assigning himself the task of showing that "*time* is that from which Dasein tacitly understands and interprets something like being at all" (BT 17/17). We understand being in a temporal way, he aims to show, because our being is grounded in *temporality* and has a distinctive temporal structure that is not immediately apparent from the way time is ordinarily understood. With regard to how philosophers have traditionally thought about time, Heidegger writes:

> This task as a whole requires that the concept of time thus gained be distinguished from the common understanding of it. The latter has become explicit in an interpretation of time which reflects the traditional concept that has persisted since Aristotle and beyond Bergson. (BT 17/18)

Heidegger soon explains why he credits Aristotle with the formulation of this "traditional concept" of time: in *Physics* IV, chapters 10–14, Aristotle explores the nature and existence of time in a way that becomes definitive for philosophy.[1] Later in the introduction, Heidegger writes, "Aristotle's treatise on time is the first detailed interpretation of this phenomenon that has come down to us. It essentially determined all the subsequent interpretations of time, including that of Bergson" (BT 25/26). The obvious question is, *why Bergson?* What is it about Bergson's interpretation of time that causes Heidegger to point in his direction, rather than Husserl's, whose lectures on time-consciousness would soon be published under his supervision? Why not Hegel, whose interpretation of time—not Bergson's—is the subject of the penultimate chapter of *Being and Time*? For that matter, why not Dilthey, whose thinking is the focus of some of Heidegger's most important early investigations into time?[2]

It is not until a footnote near the end of *Being and Time* that Heidegger offers any reasons for singling out Bergson's interpretation of time. In his discussion of the relationship between time and spirit for Hegel, Heidegger offers a brief sketch of why he considers Bergson a contemporary heir to the concept of time formulated by Aristotle.[3] The initial focus of the footnote is how Hegel's concept of time appears to be "drawn *directly* from Aristotle's *Physics*" (BT 410n/432n). Turning to Bergson, Heidegger levels the same criticism:

> Despite all differences in justification, Bergson's conception agrees with Hegel's thesis that space "is" time. Bergson just turns it around: Time (*temps*) is space. Bergson's interpretation of time,

> too, obviously grew out of an interpretation of Aristotle's treatise on time. It is not just a matter of an external literary connection that simultaneously with Bergson's *Essai sur les données immédiates de la conscience*, where the problem of *temps* and *durée* is expounded, a treatise of Bergson's appeared with the title: *Quid Aristoteles de loco senserit.*[4] With regard to the Aristotelian definition of time as ἀριθμὸς κινήσεως [*arithmos kineseos*], Bergson analyzes *number* before analyzing time. Time as space (cf. *Essai*, p. 69) is *quantitative* succession. Duration is described on the basis of a counter-orientation toward *this* concept of time as *qualitative* succession. (BT 410n/432–3n)

What makes Bergson's concept of time traditional, according to Heidegger, is its dependence on Aristotle's way of thinking about time. Bergson famously distinguishes between time and *duration* by showing that the way time is commonly represented in both thought and language is fundamentally different from the way time is *lived*. The main difference is that duration flows unceasingly and its moments permeate one another, while what we call "time" is a homogeneous medium akin to space, and its moments are juxtaposed like points or numbers. Heidegger claims that this distinction amounts to a mere reversal of Aristotle's concept of time as *arithmos kineseos*, a number related to motion, which is demonstrated by Bergson's definition of time as "quantitative succession." If time is understood as something quantitative, or something that can be counted, Bergson thinks, then it is being confused with space. This is because counting requires that whatever is counted must be numerically distinct, and such distinctness implies externality and juxtaposition, which are spatial properties. Bergson concludes that "time, understood in the sense of a medium in which we distinguish and count, is nothing but space" (TFW 91/68). The fact that Bergson developed this distinction between time and duration while he was also writing a thesis on Aristotle's concept of place is no coincidence, according to Heidegger. The signs all point back to Aristotle.

Heidegger thus portrays Bergson's understanding of time—as he also portrays Hegel's and Kant's—as fundamentally Aristotelian.[5] While it may appear to some that Bergson rethinks time radically, Heidegger contends, his reversal of Aristotle reveals that his thought remains traditional. The footnote on Hegel and Bergson continues:

> This is not the place for a critical discussion of Bergson's concept of time and other present-day interpretations of time. To

> the extent that anything essential has been gained at all beyond Aristotle and Kant, the concern is more with grasping time and "time consciousness." (BT 410n/433n)

A couple of things are interesting about this caveat: first is the way Heidegger's reference to "time consciousness" (*Zeitbewußtsein*) evokes Husserl's phenomenological analyses of time, and second is the notion that a "critical discussion" or "decisive confrontation" (*Auseinandersetzung*) with Bergson is called for. Heidegger's use of these particular terms raises more questions: Does Husserl's conception of time fall under the "other present-day interpretations" that Heidegger intends to criticize? Is a confrontation with Bergson over his interpretation of time really necessary, and if so, why? In the first edition of *Being and Time*, in a remark omitted from later editions, Heidegger's dismissal of Bergson concludes: "We shall come back to this in the first and third divisions of Part Two" (BT 410n/433n), referring to the critical interpretations of Kant and Aristotle he had originally planned for the "phenomenological destruction of the history of ontology" (BT 39–40/39–40). Although the second part of *Being and Time* was never published as such, Heidegger renews his attack on Bergson in his lecture courses immediately following its publication, reiterating and complicating the critique outlined above.[6]

In *The Basic Problems of Phenomenology*, a lecture course delivered in the summer of 1927, Heidegger recapitulates his argument that Bergson does not overcome Aristotle's concept of time. However, he goes further by accusing Bergson not only of formulating duration as a "counter-concept" to Aristotle's concept of time, but also of misinterpreting Aristotle. According to Heidegger, Bergson "does not succeed by means of this concept [of duration] in working his way through to the true phenomenon of time," yet Bergson's inquiries "are valuable because they manifest a philosophical effort to surpass the traditional concept of time" (BPP 232/329). Heidegger attacks Bergson several times in the course of an extensive discussion of Aristotle, claiming that "the Aristotelian concept of time was misunderstood in the modern period, especially by Bergson," (BPP 242/343) and that because of overly narrow interpretations of some of Aristotle's terms, "the Aristotelian definition of time remains unintelligible. Or else defective interpretations occur, for example that of Bergson, who says that time as Aristotle understands it is space" (BPP 244/345). These remarks, while not exactly fulfilling Heidegger's promise to return to Bergson in part two of *Being and Time*, still demonstrate his commitment to coming to terms with Bergson's philosophy of time.[7] Indeed, prior to criticizing Bergson,

Heidegger had provided his students with a compendium of philosophical investigations of time, remarking, "From the most recent period we may cite Bergson's investigations of the time phenomenon. They are by far the most independent" (BPP 231/328).

The same tension reappears the following year in *The Metaphysical Foundations of Logic*, a course delivered in the summer of 1928. There, he explains:

> Recently Bergson tried to conceive the concept of time more originally. He made it more clear than any previous philosopher that time is interwoven with consciousness. But the essential thing remained unresolved in Bergson, without even becoming a problem. (MFL 149/189)

Later in the course, Heidegger adds:

> Bergson first worked out the connection between a derived and an original time. But he did so in a way that went too far and said that time, once emerged, is space. . . . Bergson's analyses nonetheless belong to the most intense analyses of time that we possess." (MFL 203/262)

Although we could receive the impression from *Being and Time* alone that Bergson is more or less inconsequential for Heidegger, these lectures shortly following its publication paint a different picture. Here, Heidegger portrays Bergson's philosophy of time as comprising the most "independent" and "intense" investigations of the contemporary age. This deserves our attention, if for no other reason than because Heidegger reserves such praise for Bergson rather than for his teacher and mentor Husserl.

Heidegger presents Bergson in these lecture courses following *Being and Time* as having achieved important insights about time, yet having been unable to see what is essential. Contrary to what Heidegger's dismissal of Bergson in the margins of *Being and Time* suggests, he did not completely reject Bergson's way of thinking about time. It may appear as if Heidegger viewed Bergson merely as a cautionary example of a contemporary philosopher who attempted to rethink time but could not do so radically enough because of his dependence on Aristotle. However, a close look at Heidegger's early works shows that from his student days through the period in which he produced the earliest drafts of *Being and Time*, he already displayed both the recognition of Bergson's importance and the ambivalence about his philosophy of time that are evident in the lecture courses following *Being and Time*.

Heidegger's early encounters with Bergson help to illuminate why Bergson makes several prominent appearances in *Being and Time* only to be dismissed in a footnote near the end. Following Heidegger's path in thinking about time from his earliest writings to *Being and Time*, we can see the development of his strategy of distinguishing between "primordial time" (*ursprünglich Zeit*), or temporality, and time as we commonly understand it. Recognizing that Bergson also seeks to radically rethink time, Heidegger uses Bergson as a touchstone, returning to his thought over and over again. However, he disagrees with Bergson's account of duration as primordial time on many points. More precisely, Heidegger *comes to disagree* with Bergson on many points over the course of an engagement with his thinking that lasts more than a decade.

In what follows, I show that Heidegger appropriates certain key elements of Bergson's thinking as early as his 1915 *Habilitation* lecture "The Concept of Time in the Science of History," which contrasts the concept of time employed by natural science, particularly physics, with the concept of time needed for the study of history. Heidegger's descriptions of this contrast echo Bergson's distinction in *Time and Free Will* between the concept of time as a homogeneous medium, which he also associates with physics, and the experience of pure duration. However, Heidegger soon expresses concerns about Bergson's philosophy of life, especially with regard to the question of the appropriate method for understanding it. In an essay written in 1920, "Comments on Karl Jaspers' *Psychology of Worldviews*," and two contemporaneous lecture courses, *Basic Problems of Phenomenology* (WS 1919–20) and *Phenomenology of Intuition and Expression* (SS 1920), Heidegger takes issue with the view, which he attributes to Bergson, that life is a phenomenon that cannot be conceptually comprehended. Later, in his 1924 lecture *The Concept of Time*, Heidegger echoes Bergson again in the way he distinguishes time as it is measured by physicists and read off the clock from a more primordial experience of time. Heidegger thus plants the seeds for his interpretation of temporality in *Being and Time* by contrasting both the scientific concept of time and our ordinary understanding of time with a more "authentic" temporality of historical existence. In *History of the Concept of Time* (SS 1925), Heidegger announces a plan to make Bergson's thought the point of departure for a phenomenological destruction of the traditional concept of time that will trace it through Kant and Newton to Aristotle. In the introduction to this course, Heidegger credits Bergson with attempting to overcome the traditional concept of time by going back to a more original one, but he claims that Bergson presupposes Aristotle's concept. Finally, in *Logic: The Question of Truth* (WS 1925–6), Heidegger shows in more detail why Bergson's philosophy of time is no exception to the rule that all philosophical reflection on time in the Western tradition has been dominated by Aristotle's thought. In this course, Heidegger argues that Bergson's thinking is shaped by its "constant opposition to Aristotle's concept of time" (LQT 207/250), which becomes the theme for Heidegger's interpretation of Bergson in *Being and Time*. By showing how Heidegger deals with time at each of these stages, I hope to reveal not only how he follows in Bergson's footsteps, but also how he diverges from Bergson's path to blaze his own trail to originary temporality.

2. The Structure of the Concept of Time

The earliest trace of Heidegger's engagement with Bergson appears more than a decade before the publication of *Being and Time* in his 1915 lecture "The Concept of Time in the Science of History."[8] Heidegger's goal in this lecture is to illuminate certain structural differences between the natural sciences (*Naturwissenschaften*) and the "historical sciences" (*Geisteswissenschaften*) by means of an examination of one of their basic concepts: the concept of time.[9] His strategy is to analyze the structure of the concept of time by first examining its function in physics and then contrasting it with "the historical concept of time." He argues that because time serves a different function for historians than it does for physicists, the concept of time needed for the study of history has a fundamentally different structure than that of the one employed in attempts to determine the laws of nature. With an understanding of the difference between these two concepts of time, he proposes, "it should be possible to determine something general about the logical structure of history as a science" (S 51/418). Heidegger's project is thus to clarify the difference between the natural sciences and the historical sciences on the basis of their concepts of time.[10]

The influence of Dilthey, Husserl, and Rickert is apparent in this early piece, even though Heidegger mentions none of them by name. Somewhat less apparent is the influence of Bergson, whom he had studied in 1913 with Rickert in a course titled "Seminar on Metaphysics in Conjunction with the Writings of H. Bergson."[11] Although Heidegger also never mentions him by name, we can hear echoes of Bergson's *Time and Free Will* in Heidegger's analysis of the concept of time in the natural sciences and his descriptions of how the concept of time in history differs from it.[12] Heidegger's basic contrast between two kinds of time corresponds in several ways to Bergson's distinction in *Time and Free Will* between time as a homogeneous medium and pure duration. Indeed, Bergson's influence on Heidegger at this stage is apparent in how he characterizes both concepts of time, the natural and the historical.

The first way that Heidegger's analysis of the concept of time resembles Bergson's is in his characterization of the function of time in physics. As already mentioned, Bergson distinguishes time (*temps*)—or, more precisely, what we usually call "time"—from the duration (*durée*) of conscious life, which he considers to be fundamentally different from space. With regard to the former, Bergson writes, "time enters into the formulas of mechanics, into the calculations of the astronomer, and even of the physicist, under the form of a quantity. We measure the velocity of a movement, implying that time itself is a magnitude" (TFW 107/72). As such, the way physicists and astronomers understand time is determined by the need to measure motion. For their purposes, time is conceived as a physical quantity that can be used, most basically, to determine the speed of moving objects ($v = d/t$). Similarly, Heidegger argues that the concept of time in physics is formed according to its function in measurement:

> Motions run their course in time. What exactly does this mean? "In" time has a spatial meaning; however, time is obviously nothing spatial. . . . In the relation between motion and time, what is clearly at issue is *measurement* of motion by means of time. (S 53/421–2)

From Galileo to Einstein, as far as physics is concerned, Heidegger claims, "*It is the function of time to make measurement possible*" (S 53/422). The structure of the concept of time employed in physics can thus be described in terms of its quantitative function.

The second resemblance between Heidegger's analysis and Bergson's involves the way time is structured to serve this function. Bergson argues that operations of counting and measuring require the representation of time as a "homogeneous medium" in which terms can be separated and juxtaposed. In some respects, therefore, what physicists call "time" is closer in structure to the manifold of empty space than it is to conscious duration. According to Bergson, the lived experience of duration involves a "pure heterogeneity" of sensations, feelings, memories, thoughts, and other conscious states. Duration is thus "a succession of qualitative changes, which melt into and permeate one another . . . without any affiliation with number" (TFW 104/70). However, this is not at all the way time is understood when it serves the function of measuring motion. Whereas every moment of conscious experience is a different blend of elements, for the purpose of measurement we must be able to represent time as a sequence whose moments are the same, as a sequence of points distinguished only by their position. For Bergson, in contrast to the heterogeneity of duration, "the time which the astronomer introduces into his formulae, the time which our clocks divide into equal portions . . . must be a measurable and therefore homogeneous magnitude" (TFW 107/72). Bergson's account of time as *homogeneous* is then closely echoed in Heidegger's description of the structure of time as determined by the function of making measurement possible:

> Time presents a series with a single direction in which each point of time differs only through its place as measured by the initial point. Because one point of time differs from the preceding one only in that it follows after it, it is possible to measure time and thereby motion. As soon as time is measured—and only as time that is measurable and to be measured does it have a meaningful function in physics—we determine a "so many." The registering of the "so many" gathers into one the points of time that have until then flowed by. We as it were make a cut in the time scale, thereby destroying authentic time in its flow and allowing it to harden. The flow freezes, becomes a flat surface, and only as a flat surface can it be measured. Time becomes a homogeneous arrangement of places, a scale, a parameter. (S 54–55/424)

Although Heidegger does not employ the term "duration," his description of time as homogeneous suggests that he may be appropriating the Bergsonian distinction between *durée* and *temps* for the sake of his own contrast between historical time and the time of nature.

Not only is Heidegger's use of the term "homogeneous" suggestive here, but so is his analysis of how the concept of time in physics is derived by "destroying authentic time in its flow." According to Bergson, "science cannot deal with time and motion except on condition of first eliminating the essential and qualitative element—of time, duration, and of motion, mobility" (TFW 115/77). This challenges a deeply held assumption that time is essentially quantitative, that which is measured by the clock. If we reflect on the flow of conscious experience, Bergson argues, the moments of life are not immediately given as a "discrete" or "quantitative multiplicity," like a series of numbers or points on a line. Rather, we experience "a duration whose heterogeneous moments permeate one another" (TFW 128/85), blending together in a "qualitative multiplicity" whose elements "undergo a deep alteration as soon as we separate them from one another" (TFW 125/83). Contrary to his subsequent attack on Bergson, at this stage Heidegger also uses the categories of quantity and quality to distinguish between the concepts of time employed by physicists and historians. He explains, "The object of the science of history is human beings—not just as biological objects but rather to the extent that their achievements in the realms of mind and body actualize the idea of culture (S 56/426). Because the cultural achievements and historical events that define our present situation are in the past, Heidegger argues, the concept of time must function in such a way as to provide access to them and make it possible to interpret their meaning. That is, the concept of time must be structured so that historical facts can be established and put into context. Historians must conceive of time as divided into ages, each distinguished by its own characteristics and tendencies. Thus, Heidegger claims, "the essential element in the historical concept of time" is that "*Historical times differ qualitatively*" (S 59/431). With respect to historical dates, it may appear as if historians also deal in quantities of time, but Heidegger argues, "If I ask about the 'when' of a *historical* event, then I am asking about its place in a *qualitative historical* context, not about a 'how many' " (S 59/431).

In the historical sciences, time is understood in terms of the hermeneutical task of interpreting events rather than the theoretical one of formulating laws. Thus, with his characterization of historical time, which owes more to Dilthey than to any other thinker, Heidegger nevertheless also incorporates Bergson's views about the difference between time and duration. Like Bergson, Heidegger maintains that the time we measure and divide into discrete units, the time of the natural sciences, is fundamentally different from the qualitative, heterogeneous time of human life. This difference highlights the need for further inquiry into the time of lived experience, an inquiry that gradually points Heidegger in the direction of originary temporality.

3. Life as a Primordial Phenomenon

Although Heidegger's 1915 *Habilitation* lecture displays his earliest engagement with Bergson, his first actual references to Bergson occur in an essay and several lecture courses from 1919 to 1920, where his focus is on phenomenology as a method and life as its proper object. The works in question immediately follow the groundbreaking "The Idea of Philosophy and the Problem of Worldview" (KNS 1919), in which Heidegger presents phenomenology as a "pre-theoretical primordial science." Heidegger's aim is to show how it is possible for phenomenology to understand life philosophically without objectifying and thus distorting it. From this period until the publication of *Being and Time*, the phenomenon at the center of Heidegger's attention goes by many names, including "the primal something, life in and for itself, factic life, the historical I, the situated I, factic life experience, facticity, Dasein, being."[13] The question, for Heidegger, is how to grasp this primordial phenomenon of life without obscuring its distinctive character.

A brief but significant criticism of Bergson appears in Heidegger's 1920 essay (unpublished until 1973) "Comments on Karl Jaspers' *Psychology of Worldviews*. In this review, Heidegger presents a critical discussion of Jaspers's method and conceptual framework for understanding "psychical life" (S 76/7ff).[14] His aim, as he states early in the essay, is to "free up the real tendencies of Jaspers' work" by showing "to what extent Jaspers' approach to his tasks, his choice of methods, and his way of employing these methodological means to carry out his tasks are really in keeping with the underlying tendencies of his inquiry" (S 72/2). Heidegger thus undertakes the task of a phenomenological destruction, to draw out the "foreconceptions" (*Vorbegriffen*), or pre-theoretical notions, guiding Jaspers's philosophy. In particular, he is concerned about Jaspers's foreconception of life as a whole, as an encompassing region to be clarified from the standpoint of certain "limit-situations" or "situations of antinomy" (S 78–80/10–12).

To situate Jaspers's project with respect to other thinkers working toward a philosophy of life, Heidegger distinguishes between two meanings of "life" that tend to be conflated: "Either life is approached as the fundamental reality, and all phenomena are seen to lead back to it, so that everything and anything is understood as an objectification and manifestation 'of life,' or life is seen as the formation of culture, and this formation is thought to be carried out with reference to normative principles and values" (S 81/15). In the latter direction, life is viewed as objectifying (*Objektivieren*), "going out of itself" creatively, expressing itself through culture; and in the former direction, life is viewed as experiencing (*Erleben*), gathering in, appropriating its surroundings.[15] Heidegger applauds Jaspers for calling attention to "the problem of existence" but raises concerns about his foreconception of life as an "infinite process" that is "thought of as a thing-like object . . . *something given* in the objective medium of psychical being . . . a flowing 'stream' that bears all movements within itself" (S 84/18). If life as a whole is regarded as an unceasing flux of lived experience, would it not resist theoretical objectification? But how can a phenomenon be understood if it cannot be grasped as an object?

When Heidegger elaborates on the difficulties arising from Jaspers's foreconception of life, Bergson makes a brief appearance. It seems as if life as a whole must somehow be grasped as a theoretical object, but as Jaspers defines it, life is a process that resists objectification: "Every attempt to understand life is forced to turn the surge and flux of the aforementioned process into a static concept and thereby *destroy* the essence of life, i.e., the restlessness and movement . . . that characterize life's actualization of its ownmost qualities" (S 84/18). From this standpoint, life is a phenomenon that cannot be objectified, a process rather than a thing, and an endless process at that. We may be capable of intuiting it directly, Jaspers suggests, but we are poorly equipped to understand it intellectually. Heidegger responds:

> This characteristically Bergsonian line of argumentation suffers from a certain paralysis in a twofold sense. Apart from the fact that problems concerning meaning, concepts, and language are approached only from a very narrow perspective focusing on objective, reifying concepts, these problems are allowed to remain on the level of a very crude and vague treatment that contributes nothing to that type of treatment in which one would attempt to define the fundamental sense of life and lived experience as a whole. And instead of using this "glut on the market" to provide oneself with an air of profound philosophy (such talk of ineffability easily gives the impression that one has actually gazed upon ineffable realms), it is high time that we found genuine problems to deal with. (S 84–5/19)

It is not immediately clear what Heidegger is trying to say about Bergson here, but the remark about "objective, reifying concepts" is significant. It suggests an association of Bergson with the view, which Heidegger attributes elsewhere to Paul Natorp, that because our concepts are designed to grasp objects, they fall short when we use them to theorize about subjectivity or lived experience.[16] From this "very narrow perspective," concepts are static and therefore inadequate for grasping the dynamic constitution of life. Heidegger, on the contrary, believes that what is necessary is to develop the right kind of concepts, beginning with a more appropriate foreconception of life as a phenomenon that "is 'there' in various types of understanding and conceptual expression" (S 98/38). In other words, life needs to be regarded as a phenomenon that *objectifies itself* in a historical manner—through culture, art, religion, philosophy, and even science—and therefore lends itself to hermeneutic conceptualization.

Heidegger criticizes this "characteristically Bergsonian line of argument" in greater detail in *Phenomenology of Intuition and Expression: Theory of Philosophical Concept Formation* (SS 1920). Although Heidegger deals more extensively with Natorp and Dilthey, he mentions Bergson several times in connection with their philosophical approaches to life. Anticipating the two meanings of "life" distinguished in the Jaspers essay, Heidegger writes:

> The problematic of contemporary philosophy is centered around life as *primal phenomenon* [*Urphänomen*]: It is either that life in general is posited as the primal phenomenon and all questions are directed back to this, that is, that every objecthood is comprehended as objectivation and manifestation of life—for example the philosophies of life as they, mainly in biological fundamental orientation, are connected to the names James and Bergson, in the fundamental orientation of the human sciences to the name of Dilthey and in one that also unifies both motive groups as well as the one that is to be mentioned in what follows, to the name of Simmel. Or life is seen as culture, as manifestation, but now with a view to the fact that this culture formation and life enacts itself and is supposed to enact itself in a bond to norm-giving principles and values. The goal of such consideration of life is then a universal a priori systematics of reason as it is strived for by the Marburg school, by Rickert and in the most recent development of his ideas by Husserl. (PIE 10/15)

For his teachers Husserl and Rickert, as well as all "philosophy of culture" (*Kulturphilosophie*), the main problem is how absolutely valid claims are possible in philosophy, religion, morality, and the arts, which are historically contingent expressions of the human spirit. For Bergson, James, Natorp, and Dilthey, representing "philosophy of life" (*Lebensphilosophie*), the main problem is how lived experience can be grasped by theoretical reflection without being distorted. Taken together, what Heidegger designates as "the problem of a priori validity" and "the problem of the irrational" (PIE 20/28) comprise the "problem situation" of contemporary philosophy. In defense of phenomenology, and in an attempt to radicalize it, Heidegger argues that both sides of the contemporary problematic are out of touch with the primordial phenomenon of life.

The way Heidegger situates Bergson with respect to *Lebensphilosophie* helps to illuminate his complaints about Bergson's "narrow perspective" on concepts. Bergson belongs to the second problem group, which interprets the primal phenomenon of life in terms of *lived experience* (*Erlebnis*). For the thinkers associated with this group, philosophy itself is called into question, along with the sciences that purport to study human life: "provided that philosophy . . . should somehow be rational *knowledge*, the question arises for it whether a consideration of living experience [*Erlebens*] that does not immediately and necessarily theoretically disfigure it is possible at all" (PIE 18/25). As Heidegger explains, this concern arises in part from Kant's epistemology:

> Everything known is categorically formed material. . . . The categorically unformed, theoretically unaffected is the merely passively experienced and experience-able [*Erlebte und Erlebbare*]. Nothing can be stated about it, unless in theoretical forming, meaning at the same time, however, the demolition of the immediacy in the mediation through the intellect. Living experience is first defined as the flow [*Ablauf*] of this logically unaffected, the arising and going-along of the "I" within this flowing. . . . A knowledge as forming of this living experience qua living experience means a theoretical shaping, a logical, formally guided mediation of the unmediated immediate, respectively a rationalization of the irrational, a demolition or immobilization [*Stillstellung*] of life in the schema of concepts. (PIE 18–19/25–26)

With the foreconception of life as passively intuited, immediately given, and ineffable, the question arises of whether it is comprehensible to the intellect. If lived experience is a pure flow of conscious content without logical order, and if the function of concepts is to stabilize or immobilize the stream of experience, then how could life be grasped conceptually? For philosophers who identify life with experience unmediated by the categories of knowledge, the outcome is a kind of irrationalism.[17]

While the foregoing description of the problem of knowledge in the context of *Lebensphilosophie* faithfully articulates Bergson's motivation for pursuing a method of "philosophical intuition,"[18] Heidegger only mentions Bergson by name when he turns from the question of intuition, or access to lived experience, to that of expression:

> A second motive which Bergson particularly strongly emphasized in his "An Essay on the Immediate Data of Consciousness" (1889)

> is taken from language. Knowing communicates itself in language, in words. Language is, it is said, tailored to the spatial external world and its practical, rationally technical control. Both the meanings of words and the concepts relate to space; all logic is logic of space. . . . all theoretically conceptual apprehension of living experience, of consciousness or of the spirit is a spatialization and therefore a fundamental disfiguration. (PIE 19/26–27)

It is clear from the opening lines of *Time and Free Will* that the predominance of spatiality in human language and thought is indeed a central theme of Bergson's:

> We necessarily express ourselves by means of words and we usually think in terms of space. In other words, language demands that we establish between our ideas the same sharp and precise distinctions, the same discontinuity, as between material objects. This assimilation is useful in practical life and necessary in most of the sciences. But we may ask whether the insurmountable difficulties that certain philosophical problems raise do not come from our insistence on juxtaposing in space phenomena that do not occupy space, and whether, by setting aside the crude images around which the battle is waged, we might put an end to it. (TFW xix/vii)

We have already glimpsed how this plays out in Bergson's distinction between time, which is commonly represented as a homogeneous medium analogous to space, and duration, which cannot be understood adequately in terms of space. In fact, Bergson's concern about spatial representations motivates his strategy of showing how the problem of free will can be resolved if we think in terms of duration rather than space. The difficulty of doing so leads Bergson to prioritize the problem of method, seeking a way to think that does not treat all phenomena as if they were external objects.

Heidegger's own efforts to refine the phenomenological method run through the 1920 Jaspers essay and *Phenomenology of Intuition and Expression*, as well as his earlier courses *Phenomenology and Transcendental Philosophy of Value* (SS 1919) and *Basic Problems of Phenomenology* (WS 1919–20). Heidegger's interest in Bergson and desire to come to terms with his thought are already evident in these lectures, which represent some of his first efforts at developing the idea of phenomenology as a "primordial science" of life. In the former, while the only reference to Bergson appears in a discussion of Rickert's influence on phenomenology (TDP 136/179), Heidegger's analysis of lived experience and "the primacy of the theoretical" directly engages the concerns about grasping life as such that he calls "Bergsonian" in his Jaspers essay. He argues that the way Natorp understands the problem—and employs it in criticizing phenomenology—is that reflection turns whatever it grasps into an object, including experiences that are fundamentally subjective. Heidegger speaks of the "de-vivification" (*Entlebnis*) of lived experience (*Erlebnis*) from the theoretical standpoint: "We set the experiences *out* before us *out* of immediate experience; we intrude so to speak into the flowing stream of experiences and pull one or more of them out, we 'still the stream' as Natorp says" (TDP 78/101). While Heidegger agrees that philosophy should be concerned about theoretical objectification in its attempts to grasp life as such, he denies that Natorp, Bergson, or others in the problem group surrounding lived experience have solved the problem or even posed it correctly. He attacks the notion of experience as something "immediately given," arguing that such a thing is not pre-theoretical, but represents an "initial objectifying infringement of the environment" (TDP 69/89). The turn from objects to a pure stream of experience fails to leave the theoretical attitude for a more originary stance, according to Heidegger, but remains a de-vivification of lived experience, which is always "worldly."[19]

Heidegger resumes his attempt to turn phenomenology into a primordial science of life in *Basic Problems of Phenomenology* (WS 1919–20, not to be confused with the SS 1927 course of the same name), in which we find several significant references to Bergson.[20] As in the later *Phenomenology of Intuition and Expression*, Heidegger associates Bergson with James, Dilthey, and especially Simmel, whose latest work he characterizes as "strongly and essentially determined by Bergson" and "searching . . . for a basic intuition [*Grundanschauung*], which coincides essentially with Bergson" (BPPa 7–8/9–10; see also PIE 53/69). Anticipating a discussion of Bergson's thought that does not come to fruition in this course, Heidegger attacks the rationalistic tendency to trivialize his "partly ingenious intuition," but hastens to add, "To be sure, here I am not talking about the necessarily high estimation of Bergson made by the trendsetters, but that Bergson whom they do not know or understand in the least, but who, because of his importance, must be reckoned with in detail—through positive critical research" (BPPa 8/10, translation modified). As we will see, this is the first of several postponements of a more direct confrontation with Bergson's thought.[21]

After alluding to this important but misunderstood Bergson, Heidegger nevertheless attacks a quite familiar Bergson, the one known primarily on the basis of *Creative Evolution*.[22] His most significant remarks about Bergson occur near the beginning and the end of the course, helping to illuminate how he thinks the phenomenon of life should be conceived for it to be both accessible and expressible. For phenomenology to be a primordial science, Heidegger explains, it cannot stand outside of life and observe it, but must develop a form of research in which life itself "finds its way back into the original sources," a radical "enactment" or "actualization" (*Vollzug*) of life (BPPa 16–17/22–23). Phenomenology must therefore oppose both rationalism and irrationalism, he argues, and this requires "directing one's attitude into the vital motivations and tendencies of spirit—of the *élan vital* [vital impetus], but in a sense that is different from the mystically confused sense of Bergson" (BPPa 17/24). Heidegger thus identifies the topic of phenomenological research—the pre-theoretical, primordial phenomenon—with Bergson's *élan vital*. However, he clearly has concerns about how Bergson characterizes life, which he shares near the end of the course. There, he attacks Bergson for failing to acknowledge that life is not only dynamic and restless, but also intrinsically meaningful, "that life does not mutely flow along like a current (the way *Bergson* describes it, as a result of working through biological concepts). Rather, it is intelligible" (BPPa 174–5/231, translation modified). While Heidegger contends that philosophy should investigate "life in itself," and he recognizes how difficult it is to do so without "de-vivifying" it, or distorting its distinctive character, he rejects the notion that life is ineffable and ungraspable. Life is indeed in unceasing flux, he admits, but he denies that it is therefore beyond our grasp. Rather, life has an intentional structure, "a basic directedness in each case and always into a world" (BPPa 25/30), in terms of which it can be understood. The theoretical standpoint that objectifies life is itself an outgrowth of worldly experience, which always takes place within an already meaningful context. Heidegger insists that the primordial phenomenon of life is by no means an inarticulate, unintelligible flux, as Bergson seems to suggest, but is directed toward the world, others, and itself in definite ways that can be grasped phenomenologically.

Notwithstanding this criticism of Bergson's view of life, near the end of *Basic Problems*, Heidegger makes a comment in a very different tone about Bergson's philosophy of time. The context for his remark, which does not appear in Heidegger's notes but comes to us by way of a student transcript, is a discussion of the problem of "having oneself" (*Sich-selbst-Haben*). In the first section of the course, Heidegger distinguishes between three ways that the intentionality of "factical life" manifests itself: in an "environing-world" (*Umwelt*), a "with-world" (*Mitwelt*), and a "self-world" (*Selbstwelt*). Following this distinction, much of the course focuses on the self-world as the region to be explored by phenomenology as a primordial science of life. In opposition to the tendency to objectify the self, Heidegger says, "One should not be disappointed at not finding an 'I' in the brightness of consciousness, but rather only finding the rhythm of experience itself.—The self is present to us in the expression of the *situation*" (BPPa 195/259). The challenge for him is to understand the situation as a concrete, meaningful, "temporally particular" (*jeweilig*) expression of the self. As such, Heidegger says: "The *problem of time* is bound up with that of the situation. One can thank Bergson for the decisive accomplishment of separating the '*durée concret*' [concrete duration] from the objective 'cosmic' time'—We cannot go into that any further" (BPPa 195/259).[23] Heidegger's shift back to stressing Bergson's importance with respect to time, specifically the importance of his distinction between duration and time, contrasts sharply with his previous remarks concerning the deficiencies of Bergson's concept of life.[24]

Taking all these works from 1919–20 into view, it appears that Heidegger's worries about Bergson stem from methodological considerations surrounding the philosophy of life, specifically from the view that life is something that is ceaselessly in flux and that cannot be grasped conceptually. In the problematic of *Lebensphilosophie* as Heidegger describes it, Bergson is on the side of lived experience, tending toward irrationalism by conceiving of life as an ineffable, immediately given flux. The issue of time is not in the foreground here, yet Heidegger is advancing toward a conception of factical life as not only worldly, but also having its own distinctive temporality. He praises Bergson for distinguishing between time and duration, but criticizes him (along with Jaspers) for his view of life as an endless process that cannot be grasped theoretically. Interestingly, though, Heidegger's own concerns about the primacy of the theoretical echo some of Bergson's (and Natorp's) worries about whether life can be understood philosophically without being immobilized, objectified, or spatialized. Even so, in thinking about life as a "primordial phenomenon," Heidegger distances himself from the view that any attempt to grasp life necessarily distorts it because of the static character of concepts and the spatializing tendencies of language and thought. Nevertheless, when Heidegger returns to the question of time in 1924–5, he again distinguishes, as Bergson does, between the quantitative, homogeneous time of the natural sciences and the qualitative, heterogeneous time of history and historical beings.

4. To Understand Time in Terms of Time

In his earliest investigation into time, Heidegger had argued that understanding history requires a concept of time that differs fundamentally from the one employed in the natural sciences, which immobilizes what we experience as a continuous flow. However, through his efforts to bring phenomenology into its own as a primordial science of life, he had come to believe that the notion of time as a flowing stream is also determined by the theoretical attitude. Even the stream of lived experience is an objectification of a peculiar sort, contrary to what champions of life such as Jaspers, Natorp, James, and Bergson seem to think. Because they regard life as psychological or biological, their access to the phenomenon of *factical* life is limited, in Heidegger's view. Life is essentially historical, he maintains, insofar as it happens in a temporally particular situation, embedded in a world and a tradition. Having argued that phenomenology needs to confront life and *Lebensphilosophie* in order to be more radical, Heidegger returns to the question of time. Now, though, it is no longer a matter of clarifying the difference between the natural and historical concepts of time, but rather of showing how both are rooted in the kind of *being* that belongs to history as well as nature.

In the now famous 1924 lecture "The Concept of Time," delivered to the Marburg Theological Society, Heidegger distinguishes all scientific understanding of time from our everyday, pre-theoretical experience of it, and then he describes a less common but more "authentic" temporality in which he thinks both originate. Following the lead of Aristotle and Augustine, he seeks to show how time can be, on the one hand, "that within which events take place" (CT 3/109), something encountered along with change or motion, something that can be measured, and on the other hand, something that depends on the mind or spirit, a "disposition" of the being that measures it. Rather than attempting to define time, however, he inquires about the presuppositions of the concept of time employed by the natural sciences and the ordinary understanding of time as a measurable quantity. The brief analysis that Heidegger presents in this lecture is, in many ways, a sketch of the phenomenological interpretation of time that eventually appears in division two of *Being and Time*, where he attempts to show how both the time of nature and the time of history are grounded in the temporality of Dasein.[25] The influences of Bergson, Husserl, and Dilthey are all evident in this lecture, even though Heidegger mentions none of them by name. We can see more clearly how he follows Bergson in the way that his analysis proceeds from how physicists conceive of time to how we experience it in the anticipation of death.

Heidegger begins with the straightforward, traditional question "What is time?" and explains why he must resist the temptation to take eternity as a starting point, as he thinks a theologian might. "If the philosopher asks about time," he says, "then he has resolved *to understand time in terms of time*" (CT 1/107). He confesses, however, that his treatment of time will not be strictly philosophical either, because he aims not to offer a universal definition of time, but to clarify how it shows up in everyday life, prior to scientific and philosophical investigation. His primary concern is to shed light on the experiences of time presupposed by current scientific understanding. In at least one respect, Heidegger argues, Einstein's theory of relativity is not fundamentally different from Aristotle's physics: for both, time is "that within which events take place" (CT 3/109), "that within which things change" (CT 4/109), and ultimately that within which duration can be measured. Understood in this way, time is a medium in which natural events occur and can be quantified. But is this really the most fundamental phenomenon of time? Returning to the topic of his 1915 essay on time, Heidegger asks: "How does the physicist encounter time?" (CT 4/109). For the phenomenologist, this question signals a shift from asking about *what* time is to asking about it *how* it is approached, apprehended, and determined.

Heidegger's answer, that physicists encounter time first and foremost by *measuring* it, leads to a phenomenological analysis of the way time is measured with clocks. He explains:

> A clock shows the time. A clock is a physical system in which an identical temporal sequence is constantly repeated, with the provision that this physical system is not subject to change through any external influence. The repetition is cyclical. Each period has an identical temporal duration. . . . The clock measures time in so far as the stretch of the duration of an occurrence is compared with identical sequences on the clock and can thereby be numerically determined. (CT 4/110)[26]

Because the length of each unit of measurement may be arbitrary as long as it remains constant and recurring, it does not matter when an event whose duration is to be measured begins or ends. What matters is that the event's beginning and ending can each be assigned a position, or "now-point," and the distance between them quantified. Understood as such, each point of time is earlier than some and later than others, and each point is indifferent to the events taking place. What do the indifference of the point and the arbitrariness of the unit reveal about time? Heidegger echoes Bergson again: "This time is thoroughly uniform, homogeneous. Only in so far as time is homogeneous is it measurable" (CT 4–5/110).

Heidegger goes on to highlight the difference between how time shows itself when measured and how we encounter it prior to measurement. He claims that although we use clocks to measure quantities of time, "[w]hat primarily the clock does in each case is not to indicate the how-long or how-much of time in its present flowing, but to determine the specific fixing of the now" (CT 5/110). While a clock makes it possible to assign a number to the "now," it can be encountered and expressed without clocks merely in relation to regularly recurring changes in the environment, such as the rising and setting of the sun. This causes Heidegger to ask, "Is the now at my disposal? . . . Am I myself the now and my existence time?" (CT 5/111). There is a sense in which time is not just something we measure, but something already felt and understood prior to the use of a clock, something about the mind or spirit. Time, in this sense, is a property not of nature, but of Dasein, "that entity in its Being which we know of as human life; this entity in the *specificity* [*Jeweiligkeit*] of its Being, the entity that we each ourselves are, which each of us finds in the fundamental assertion: I am" (CT 6/112).[27] The capacity to say "now," to tell time by the sun, and to read time off a clock distinguishes us from anything merely "within time." As Heidegger puts it, Dasein "is in time in a distinctive sense" (CT 7/112), or better yet, Dasein "*is time itself*, not *in* time" (CT 14/118).[28]

With to the turn to Dasein, Heidegger seeks to understand time ontologically, as not just a concept but as a way of being. What does it mean to say that Dasein *is* time? First, Dasein is its past, in a sense that is clear from the way it relates to death. This can be seen most clearly, for Heidegger, in our "anticipation" or "running ahead" (*Vorlaufen*) toward the possibility of our own death. Being finite, we anticipate a time when we no longer exist, when our life is a thing of the past. However, contrary to conventional wisdom, he claims, "This past is not a 'what,' but a 'how,' indeed the authentic 'how' of my Dasein" (CT 12/116). Dasein *is* its past insofar as it is temporal and historical, and "[t]he past [*Vergangenheit*]—experienced as authentic historicity—is anything but what is past [*Vorbei*]. It is something to which I can return again and again" (CT 19/123). Likewise, Heidegger argues, "[i]n running ahead Dasein *is* its future, in such a way that in this being futural it comes back to its past and present" (CT 13/118). The future, in this sense, is "the authentic 'how' of being temporal, that way of Being of Dasein in which and out of which it gives itself time," thus it is "*the fundamental phenomenon of time*" (CT 13–14/118). Finally, in contrast to the "bad present of the everyday" there is an authentic present in which "[r]unning ahead seizes the past as the authentic possibility of every moment of insight [*Augenblick*], as what is now certain" (CT 15/119). The anticipation of death thus brings to light a phenomenon of time, "authentic temporality," which is a distinctive unity of the past, present, and future that for Heidegger is constitutive of Dasein. As opposed to the familiar time of everyday life, in which the "now" is primary, in authentic temporality there is a primacy of the future.

This authentic temporality is covered up, for Heidegger, by the measurement of time. In reckoning with time, measuring it by the clock, and calculating quantities, we still encounter the time that Dasein *is*, but in modified form. In everyday life, when we are busy taking care of ourselves and others, our concern with time is directed toward *how long* things last and *how much* time we have—that is, time as a quantity. For Heidegger, this way of dealing with time is indicative of inauthentic temporality, in which the present is primary rather than the future. The measurement of time discloses time as if it were something present together with the things around us. "To bring time into the 'how much' means to take it as the now of the present. To ask after the 'how much' of time means to become absorbed in concern with some 'what' that is present" (CT 15–16/119). For the purposes of everyday life, we tend to understand time as an entity, a "what" that can be measured, rather than a way of being, how we come back to the past in anticipating the future. The future, as we ordinarily understand it, is something calculable, "not the authentic, futural being of the past, but the future that the present itself cultivates for itself as its own" (CT 16/120). Absorption in the present makes the past inauthentic as well: the past [*Vergangenheit*] is what is past [*Vorbei*], it is irretrievable" (CT 19/122),[29] and history is "an irretrievable busyness" (CT 19/122), another "what" subject to observation, as opposed to a "how" of Dasein. For the ordinary way of understanding time, "the νυν [now] is the μέτρον [measure] of past and future. Then time is already interpreted as present, past is interpreted as no-longer-present, future as indeterminate not-yet-present: past is irretrievable, future indeterminate" (CT 17/121). This view of time as a quantity is geared to the clock, by means of which time is always accessible as "now." The crucial difference, for Heidegger, is between the mode of temporality for which the future is primary—the "authentic"—and the mode characterized by absorption in the present—the "inauthentic."

The primacy of the present clarifies what it means to be "in time." If temporality is the time that Dasein *is*, then what is its relationship to time as that *in which* everything happens? According to Heidegger, this ordinary understanding of time "as a sequence constantly rolling through the now . . . out of an infinite future into an irretrievable past" (CT 18/121) is profoundly different from authentic temporality, which is covered up by the primacy of the present. The concept of time that Heidegger attributes to the entire philosophical tradition from Aristotle to Bergson is, in his view, derived from *in*authentic temporality.[30] On the traditional interpretation, time is irreversible and homogeneous, composed of a succession of "now-points" with two fundamental characteristics: it is irreversible and homogeneous. Heidegger interprets the irreversibility of time as the sign of a previous reversal: from the direction of anticipation, "running ahead" to the past from the future, to the direction of running through the present into the past. The homogeneity of time, he argues, derives from an operation like the one described by Bergson: "*Homogenization* is an assimilation of time to space, to Presence [*Präesenz*] pure and simple; it is the tendency to expel all time from itself into a present. Time becomes fully mathematized, becomes the coordinate *t* alongside the spatial coordinates *x*, *y*, *z*" (CT 18/121–2). While this homogenization of time is obvious in the sciences, it also determines the understanding of time most common in our everyday lives.[31] Heidegger concludes that the challenge is to rethink time, as well as history, in terms of the temporality of Dasein, which "is futural by running ahead to the certain yet indeterminate past" (CT 20/123). This becomes the project, first, of an article on Dilthey with the same title as Heidegger's lecture, and ultimately of division two of *Being and Time*.[32]

Although Heidegger neither mentions Bergson explicitly in this lecture nor employs the concept of duration, he follows Bergson's path in a number of crucial ways. First, Heidegger approaches temporality through the scientific understanding of time and its presuppositions. More precisely, he gains access to the phenomenon by investigating how physicists encounter time by measuring it. This leads Heidegger to an analysis of clock usage, through which he concludes that clocks can do no more than indicate "now" and that measuring or counting time depends on a disposition of the mind. Whereas Bergson argues that measurement deals with a spatialized version of time, which symbolizes duration as a homogeneous medium in which things occur, Heidegger considers how the time that is measured—the "what" or "how much"—is related to a more primordial phenomenon of time, authentic temporality. While there is a great deal of difference between Heidegger's conception of the origin of time and Bergson's, particularly Heidegger's view of the primacy of the future in authentic temporality, the fundamental distinction between the pre-theoretical experience and the concept of time—the "originary" and the "derived," as Heidegger later calls them—is a Bergsonian theme.[33] Moreover, Heidegger appropriates Bergson's critique of the concept of time as homogeneous and measurable for obscuring our view of a more authentic kind of time. Finally, Heidegger follows Bergson by avoiding the reification of time, replacing the question "What is time?" with questions about how it is lived. Like Bergson, who argues that a more fundamental experience of time is distorted by the concept of it, Heidegger maintains that originary temporality at once shows through and is covered over by the concept of time. This creates a need to "talk temporally about time" and to "repeat temporally the question of what time is" (CT 22/124). Bergson's effort to do the same accounts for the leading role Heidegger gives him the following year in his outline for a phenomenological destruction of the history of the concept of time.

5. The Time That We Ourselves Are

If "The Concept of Time" shows how profoundly time and authentic temporality differ for Heidegger, then *History of the Concept of Time* (SS 1925) clears the way for that distinction to be worked out in more detail on the basis of a phenomenological analysis of Dasein and a historical destruction of the traditional concept of time. In the introduction to this course, Heidegger outlines the project that will soon become *Being and Time*, then conducts a sympathetic critique of phenomenology and a preparatory analysis of Dasein that would be expanded into division one of *Being and Time*.[34] This outline reveals just how pivotal Bergson is for Heidegger at this point with respect to the history of philosophical attempts to understand time. Heidegger's plan is to begin the course with a historical and critical introduction to phenomenology, followed by an analysis of the phenomenon of time and the derivation of the concept (first part), a history of the concept of time (second part), and finally an interpretation of temporality as the horizon for being in general and for history and nature as regions of being (third part). Heidegger clearly affirms his commitment to coming to terms with Bergson's thought in his brief remarks about the part of his project that will deal with the history of philosophy:

> The *Second Part*, "The History of the Concept of Time," begins in the present and works backwards: *First*: Bergson's theory of time. *Second*: The concept of time in Kant and Newton. *Third*: The initial conceptual discovery of time in Aristotle. Why these three major stages of the history of the concept of time are explored will become evident from the investigation itself, inasmuch as these three stages constitute stations at which a relative transformation of the concept of time has occurred. (HCT 8–9/11)[35]

This indicates that Heidegger is planning a serious critical engagement with Bergson's "theory of time," which represents a "relative transformation" of the concept following those of Newton and Kant. Once he has shown how the concept of time is derived from the temporality of Dasein, the being that "is always we ourselves" (BT 41/41), Bergson will occupy the center of his attention. However, Heidegger continues:

> I say "relative" because basically the concept of time as Aristotle conceived it is retained throughout. Bergson in fact makes an attempt to go beyond this concept to a more original one. This justifies our treating him separately within the question of the historical concept of time. Basically, when we consider the categorial fundaments which he presupposes, namely, quality and succession, Bergson does not advance the matters at issue and so remains traditional. (HCT 9/11–12)

Here, then, Heidegger not only acknowledges Bergson's importance with respect to time, but he also once again reveals a certain ambivalence. On the one hand, Bergson both recognizes the need to overcome the traditional concept of time and makes a genuine effort to uncover the "original" phenomenon from which it is derived, which he calls "duration." But, on the other hand, Bergson's access to this phenomenon is blocked by his presuppositions about time, particularly those that reveal his dependence on Aristotle's thinking. Anticipating his critique of Bergson in *Being and Time*, Heidegger thus claims that Bergson's attempt to think more radically about time fails to escape the confines of the Western philosophical tradition. Heidegger will not attempt to demonstrate this, however, until he has first shown how the concept of time is derived from the temporality of Dasein.

Contrary to what the title of the course suggests, *History of the Concept of Time* deals only very briefly with the issue of time, with Heidegger postponing a more substantive discussion to show how access to the phenomenon of temporality must be secured. Going back to the topic of his 1915 *Habilitation* lecture, Heidegger approaches the difference between the natural and historical sciences from a fresh perspective, replacing the epistemological question of how time is conceived by each with the ontological question that sets the stage for *Being and Time*. Now the guiding question for Heidegger is how history and nature, interpreted as regions of being, have come to be understood with time as a "horizon" of intelligibility. While historians focus on establishing a chronological order, and natural scientists are concerned with the measurement of time, "both historical reality and natural reality are continuities that run their course *in time* and are traditionally understood as such. . . . Viewed simply from the outside, history and nature are temporal" (HCT 5/7). In this respect, at least, there is no difference between historical events and natural processes. Yet there is another crucial distinction here:

> To the totality of temporal reality we tend to juxtapose the *extratemporal* constituents which, for example, are the topic of research in mathematics. In addition to these extratemporal constituents in mathematics we are familiar with *supratemporal* constituents in metaphysics or theology, understood as eternity. In a very schematic and crude way, time already announces itself as *one* "index" for the differentiation and delimitation of domains of being as such. (HCT 5/7–8)

Granted that time is conceived differently according to its function in the research carried out by these disciplines, historians and natural scientists alike deal with entities that are *within* time—but what exactly does that mean? The difference between being in time and beyond it may seem self-evident, but why does the concept of time perform this ontological function of distinguishing between these domains of reality, the temporal and the timeless? These questions demand further investigation into the traditional concept of time and its origin, Heidegger argues, but first the method and thematic field of inquiry need to be secured. Consequently, he devotes *History of the Concept of Time* to a detailed history and critique of phenomenology, followed by a detailed phenomenological analysis of Dasein, in which the question of time remains only in the background.

Beyond the outline for the course, the scattered references to Bergson in *History of the Concept of Time* are largely in keeping with remarks from Heidegger's lecture courses of 1919–20, where Bergson's name often appears in association with other contemporary thinkers from whom Heidegger wishes to distinguish himself. Heidegger first mentions Bergson alongside James, Dilthey, and Husserl in a discussion of the "origin and breakthrough" of phenomenology in the late nineteenth century, remarking vaguely that "[Bergson's] theory of the immediate data of consciousness . . . goes back to the ideas of Brentano's psychology" (HCT 23/28).[36] Here, for the first time, Heidegger connects Bergson not just to *Lebensphilosophie*, but to phenomenology in its initial phase of distinguishing itself from empirical psychology. Shortly thereafter, however, Heidegger offers this disclaimer with regard to Husserl's discovery of categorical intuition: "Intuition in the phenomenological sense implies no special capacity, no exceptional way of transposing oneself into otherwise closed domains and depths of the world, not even the kind of intuition employed by Bergson" (HCT 47/64). In Heidegger's discussion of the early development of phenomenology in the first quarter of the twentieth century, he notes Bergson's influence on Max Scheler and makes another vague, somewhat puzzling remark linking Bergson's philosophy with Husserl's:

> Also in this period [1913–14], Bergson gradually became known in Germany. This was basically due to Scheler, who recognized Bergson and his significance quite early, and then was influenced by him in return. Scheler was instrumental in having Bergson translated into German. This recognition of Bergson also brought, within Husserl's work, the *investigations of internal time consciousness*. (HCT 92/126)[37]

Although this anticipates Heidegger's remark in *Being and Time* that for Bergson, "the concern is more with grasping time and 'time consciousness' " (BT 410n/433n), Heidegger does not clarify what he thinks Bergson's analyses of time have in common with Husserl's phenomenological investigations. Later in the discussion, however, Heidegger says of Scheler, "because he is strongly influenced by Bergson and Dilthey within the traditional way of questioning, he comes closer to the question which concerns us" (HCT 126/174), namely the question of being of that entity called "man" or "the person." While criticizing Scheler's interpretation of personhood for its ontological inadequacy, Heidegger nevertheless says that Scheler "makes some essential progress on the question of the relationship of the animate and psychic to corporeality," such that "in this question, Scheler, under the influence of Bergson, has made the furthest advances to date" (HCT 128/177). So here again we see a marked ambivalence toward Bergsonism, which Heidegger now connects not only with Dilthey's philosophy of life, but also with Scheler's and even Husserl's phenomenology.

Despite the positive influence that Heidegger attributes to Bergson, it appears that his purpose for bringing Bergson in closer proximity with phenomenology is to suggest that Bergson suffers from the same "fundamental neglect" of ontology as Dilthey, Husserl, and Scheler. As in previous courses, Heidegger makes an effort in *History of the Concept of Time* to radicalize phenomenology by showing what is necessary for it to gain access and give expression to "the things themselves." After detailing the breakthroughs of phenomenological research, he argues that phenomenology is not faithful to its principle insofar as its "thematic field of investigation" is ontologically obscure. He thus develops an "immanent critique" in a double sense: first, it operates from *within* phenomenology, aiming to transform it, and second, it is a critique of *immanence*, the pure, absolute, character of consciousness yielded by the phenomenological reductions. Heidegger's critique takes aim at the method of reducing experience to a sphere of immanence, bracketing the reality of things beyond consciousness (the transcendental reduction) and the individuality of any factical, existing consciousness (the eidetic reduction).[38] These reductions are designed to lead the phenomenologist from the natural attitude, according to which all our experiences can be explained as processes in the natural world, to "the region of *pure lived experiences*, of *pure consciousness with its pure correlates*, the region of the *pure ego*" (HCT 96/131). This region is a field of pure phenomena, a "stream of lived experiences," that can be reflected upon and described strictly in terms of their intentional structures. Heidegger's main concern is that Husserl demonstrates how to bring this region into view but fails to "raise the question of the *being of this region*, of the *being of consciousness*" (HCT 102/140). While Husserl draws a "fundamental distinction" between pure consciousness and the reality that transcends it, according to Heidegger, "not once is a question raised regarding the kind of being which the distinguished members have, or the kind of being which consciousness has" (HCT 114/157–8). Consequently, for Heidegger, phenomenology "operates in a *fundamental neglect*" (HCT 115/158) of ontology, a neglect which reveals that "*even phenomenological research stands under the constraints of an old tradition*" (HCT 128/178). Heidegger still maintains that phenomenology is the most promising research program in philosophy, but he believes that it cannot fulfill its promise without clarifying the being of consciousness and the meaning of being in general.

Although it is not immediately evident where Bergson stands with respect to this critique of phenomenology, Heidegger's description of "personalistic psychology" in Dilthey, Husserl, and Scheler is illuminating. One of the ways that Dilthey influenced phenomenology, according to Heidegger, was in his strong opposition to naturalistic psychology and his efforts to develop a philosophy of the human being that "does not take him for its object as a thing of nature . . . but instead *understands* him as a *living person*" (HCT 117/161). Husserl, also in opposition to naturalism, distinguishes between natural objects and the immanent "unity formations" of personality, character, soul, or spirit.[39] In his studies of the constitution of the natural and human worlds, Husserl describes a "naturalistic attitude" and contrasts it with a personalistic one that does not reduce consciousness to natural processes.[40] Finally, Scheler emphasizes "the special character of being a person," which he defines as a "performer of acts" and insists cannot be understood correctly if the person is reified, that is, "regarded as a *thing* or *substance*" (HCT 126/175).[41] Although Scheler, "under the influence of Bergson," makes important advances, according to Heidegger, his thought nevertheless suffers from a neglect of ontology that is apparent in his silence about the being of this performer. Presumably, Heidegger would say the same sort of thing about Bergson's efforts in *Time and Free Will* to dispel the confusion of a living self with a static object: it is not enough to deny that persons are things without providing a positive account of the kind of being that belongs to those entities we call "persons."

Heidegger's own efforts to remedy the fundamental neglect of ontology, which he criticizes especially in Husserl and Scheler, lead him back to the structures of Dasein and the different experiences of time sketched in "The Concept of Time." Although there are no further references to Bergson in *History of the Concept of Time*, through Heidegger's "modification" of the thematic field of phenomenology it becomes clearer how his previous analyses of time will feature in his ontology. Because time has traditionally been used as a basis for distinguishing between kinds of being (temporal, supratemporal, extratemporal), Heidegger proposes to use it as a "guiding clue" in pursuit of the meaning of being. Beginning, as he will again in *Being and Time*, with the question of being and the entity who asks it, he defines Dasein as fundamentally temporal and historical. Dasein is "*an entity which is in the to-be-at-its-time* [*Jeweilig-es-zu-sein*]" (HCT 153/206), so that it can only be adequately understood in its "whileness" or "temporal particularity" (*Jeweiligkeit*). All the ontological structures of Dasein must be regarded in the light of this fundamental characteristic, Heidegger argues, and thus in the light of time. Accordingly, he examines Dasein in terms of the time that is most familiar, "everydayness" (*Alltäglichkeit*), and the way Dasein shows itself in everydayness. Time thus illuminates Heidegger's analysis of being-in-the-world here in a way that becomes less obvious in *Being and Time*: "This in-being, which we now should never understand in a primarily local and spatial sense, is rather, as being-involved-with, defined by temporal particularity [*Jeweiligkeit*]" (HCT 159/213).[42] Even though he intentionally postpones further treatment of the concept of time until after his exposition of Dasein, temporality provides both the initial and final determination of this being. In the final moments of the course, having argued that Dasein is authentically itself only in the anticipation of death, Heidegger writes:

> *Not "time is" but "Dasein qua time temporalizes its being."* Time is not something which is found outside somewhere as a framework for world events. Time is even less something which whirs away

> inside consciousness. It is rather that which makes possible the being-ahead-of-itself-already-being-involved-in, that is, which makes possible the being of care. (HCT 319–20/442)

What Heidegger describes here is neither the time of nature nor even the time of history, but "the time which we ourselves are" (HCT 320/442), the *temporality* that distinguishes human beings from anything that occurs "in time." Because the course comes to an end with these remarks, never having reached the traditional concept of time, his students would have to wait to hear more about how this originary temporality differs from Bergsonian duration.

6. A More Original Concept of Time

The last lecture course Heidegger delivers prior to writing of *Being and Time* contains the most detailed discussion of Bergson in his work from this period. In *Logic: The Question of Truth* (WS 1925–6), Heidegger experiments with developing a "phenomenological chronology," an inquiry into the "ur-temporality" (*Temporalität*) of the basic concepts of logic (LQT 169/199).[43] The second part of this course resumes the project outlined in *History of the Concept of Time*, and it is devoted more directly to the task of understanding time than that course turned out to be. It includes a detailed interpretation of Kant's concept of time, preceded by shorter discussions of Hegel's interpretation of time in the *Philosophy of Nature* and Aristotle's influence on Hegel and Bergson.[44] This part of the course follows up on Heidegger's suggestion that Bergson attempts to go beyond the traditional concept of time but ultimately "does not advance the matters at issue." Heidegger argues, as he does in *Being and Time*, that Bergson's concept of duration is a mere reversal of the concept of time developed by Aristotle in *Physics* IV. He criticizes both Hegel and Bergson for deriving their concepts of time from the *Physics*, proving that "right up to today, time has always been understood as now-time (*Jetzt-Zeit*)" (LQT 204/246). If he can show that two of the most apparently radical, innovative investigations of time in recent years are still confined by traditional thinking, then this will provide the motivation for his own inquiry into temporality.

Two separate discussions of Bergson appear in *Logic: The Question of Truth*, and together they provide the background for Heidegger's dismissal of Bergson in *Being and Time*. In the first discussion (§19), Heidegger offers a brief interpretation of Bergson's concept of duration and explains how it is derived from Aristotle. Here, he emphasizes that "Bergson is guided by a correct instinct" (LQT 207/249), but he maintains that a critical discussion is necessary because "current opinion takes it as obvious that Bergson has discovered a new concept of time (LQT 208/251). His goal is to show, on the contrary, that "Bergson falls into the very concept of time that he is trying to overcome" (LQT 207/249). The second discussion (§21) follows a section devoted to tracing Hegel's interpretation of time back to Aristotle's and showing how, in the way he connects time with space, Hegel passes over temporality. Returning to Bergson, Heidegger attacks "Bergson's thesis (that time is space)" (LQT 221/266) with reference to two short passages from *Time and Free Will*, one linking time and space, the other linking space and number. Although Heidegger registers a number of complaints with Bergson's attempt to overcome the traditional concept of time, his two primary concerns are with the thesis "time is space" and the definition of duration as "qualitative succession." In Heidegger's view, these aspects of his thinking show that "Bergson's path for arriving at his thesis that time is space is different from Hegel's, but in principle he coincides with Hegel" (LQT 223/268). Consequently, although Bergson's philosophy of time is "independent" and "appears to attain some new insights," it is "no exception to the rule" (LQT 206/249) that all philosophers, even contemporary ones, understand time in the way that Aristotle defined it.

How does Bergson attempt to overcome the traditional concept of time? In the first place, he distinguishes between time and duration, with duration signifying "lived time." The problem, according to Heidegger, is that Bergson defines duration by simply taking the traditional concept of time and turning it on its head. Heidegger writes:

> A clear indication that Bergson failed to break through to a conception and categorical knowledge of primordial time is the fact that he understood even lived time—duration—as "*succession*," with the sole proviso that the *succession* of lived time is, he says, not a quantitative *succession* laid out in individual now-points. Instead, this *succession* is a qualitative one, in which the individual moments of time—past, present, and future—permeate each other. (LQT 207/249–50, t. M.)[45]

Because Bergson thinks of duration as a different kind of succession than time, but a succession nonetheless, Heidegger suggests, his thought remains determined by the traditional concept of time. This concept, which Heidegger calls "now-time" both here and in *Being and Time*, formalizes the "ordinary understanding of time" (*vulgären Zeitverständnis*) (LQT 204/246), which in turn arises from the "everyday experience of time" (*alltägliche Zeiterfahrung*). More specifically, this understanding arises out of the experience of taking care of and reckoning with worldly events, which we regard, just as we do all historical occurrences and natural processes, as happening "within time." This pre-theoretical orientation determines how philosophers, beginning with the ancient Greeks and continuing up to the present, conceptualize time: "Wherever time is seen in connection with world, nature, or created beings, *it is understood as now-time, and 'temporal' means: occurring and running its course 'within time'* (LQT 206/249, emphasis added).[46] If, Heidegger suggests, one of the essential characteristics of ordinary time is succession, then Bergson's concept of duration as a "qualitative succession" does not break through to the primordial phenomenon of time.

Not only does Bergson's reference to succession indicate that his concept of duration is fundamentally traditional, in Heidegger's view, but his replacement of quantitative with qualitative also reveals the kinship of his thought with Aristotle's. It is well known that quality and quantity are Aristotelian categories, and corresponding to those categories, the distinction between duration as "qualitative succession" and time as "quantitative succession" suggests to Heidegger that Aristotle's definition of time is the point of departure for Bergson's thinking. In a passage that reappears with some revisions in *Being and Time*, Heidegger writes:

> [E]ven though it is not completely obvious, Bergson conducts his analysis of time in constant opposition to Aristotle's concept of time. In his early years Bergson made extensive studies of Aristotle, and in 1889, the same year his first investigation on time and duration appeared, he published a small treatise, his dissertation, *On the Concept of Place in Aristotle*. Place and time are treated by Aristotle in the same book. And it is quite clear how Bergson arrived at his concept of duration in opposition to quantitative time, namely, by misunderstanding Aristotle's definition of time as ἀριθμὸς κινήσεως [*arithmos kineseos*], i.e., as the number—or better, the what-gets-numbered [*Gezählten*]—of movement. (LQT 207/250)[47]

Heidegger thus offers two pieces of evidence in support of Bergson's dependence on Aristotle. The first is the publication of an essay on Aristotle's concept of place in the same year as Bergson's *Time and Free Will*. The second is Bergson's replacement of time as a quantitative succession, which is implied by Aristotle's definition of time as something numbered (or counted) with respect to motion, with his own concept of duration as qualitative succession. The first piece of evidence is interesting, but it leaves completely open the question of how these two early works, one focusing on Aristotle and the other making almost no mention of him, are related. The second piece of evidence is more compelling, and Heidegger explores it in greater depth in his second discussion of Bergson.[48]

Heidegger begins the second discussion with the following quote from *Time and Free Will*: "Time, understood as a field in which we make distinctions and count, is nothing else but space" (LQT 221/266–7).[49] This passage is the only one Heidegger cites to support his interpretation that, according to Bergson, "time is space." On the one hand, this is incorrect, Heidegger argues: "Time is not space, any more than space is time. But time is simply the possibility wherein the being of space can be determined existentially-temporally—but not because the being of time is space, but because being in general, as the being of every being, must be conceived in terms of time" (LQT 222/267). Heidegger suggests a reversal of his own; the problem is not that the being of time is understood in terms of space, but that the being of space—and being in general—is understood in terms of time. Thus, Heidegger anticipates his own thesis in *Being and Time* that time is the horizon for any understanding of being, although here he specifies that "the *human* understanding of beings . . . is possible from out of time" (LQT 222/267).[50]

On the other hand, Bergson's thesis offers further evidence that his understanding of time is based on Aristotle, and even worse, "on an inadequate analysis and interpretation of Aristotle's definition of time" (LQT 222/268). For what leads Bergson to claim that time is—or, more precisely, *is confused with*—space has to do with the relationship between number and space. Heidegger explains:

> Characteristically, Bergson prefaces his analysis of time with an analysis of number, that is to say, he orients his entire investigation to Aristotle. But Bergson himself understands number in terms of space. He says . . . "every clear idea of number implies an intuition of space." (LQT 223/268)

On Heidegger's reading, this second quotation from *Time and Free Will* points back to Aristotle's concept of time as an *arithmos kineseos*, a number related to motion. When Aristotle defines time in this way, Bergson thinks he is reducing it to a manifold of points juxtaposed in space, which are counted as a moving thing traverses them. However, nows cannot be counted like spatial points because "in an act of counting, a now that is counted in this way—a now of time—could not sustain itself so as to make it connect with another now, i.e., so as to be taken collectively, along with that other now, without remaining simultaneous with that other now" (LQT 223/268). In other words, lived time, in which moments flow into one another, can only be measured or counted once those moments are separated and distinguished. But to separate them is to treat them as if they were given simultaneously, which is to translate temporal succession into spatial juxtaposition. Therefore, to define time as Aristotle does is to reduce time to space. This is Heidegger's explanation for how Bergson's thesis that time is space is based on Aristotle's definition.

In the course of these two discussions of Bergson, Heidegger lodges several other complaints in addition to the primary criticism that he fails to go beyond the traditional concept of time. With respect to duration, Bergson presents it as "lived time," which is "merely object-time or world-time, insofar as it is considered in the way it shows up in consciousness" (LQT 207/249). Another problem is that Bergson distinguishes between time and duration on the basis of the difference between quantity and quality, yet "he provides no principled discussion of these two guiding threads, but simply presupposes them as already known. Moreover, he describes qualitative time—duration—merely in images, with not a word about working out any kind of concept" (LQT 207/250). Likewise, Bergson characterizes duration as "real time," but "he says nothing about the meaning of 'reality,' and tells us nothing about the ontological nature of the life or consciousness wherein he finds real or lived time" (LQT 222/267). Finally, by identifying time with space, Bergson "fails to understand time in its own sense" (LQT 207/250); that is, he fails to recognize that the kind of time he distinguishes from duration "is precisely time—however, only a specific temporal mode of time" (LQT 223/268). Heidegger's complaints anticipate remarks that appear in *Being and Time* about Bergson's interpretation of time being "ontologically indeterminate and insufficient" and passing over the "genuine phenomenon of time" at the root of all our talk about things happening "within time."[51]

These discussions form the backdrop for Heidegger's dismissal of Bergson in *Being and Time*. While they are relatively brief, they offer a fairly clear picture of why Heidegger acknowledges Bergson as a pivotal thinker with respect to time yet describes his attempt to radically rethink time as a failure. What is important about Bergson's thought is not its novelty, for Heidegger more or less denies that there is anything truly new in his interpretation of time. But the "correct instinct" that Heidegger praises near the beginning of the discussion is the one that leads Bergson to seek "a more original concept of time." As a whole, Heidegger's critique is designed to show that Bergson's way of doing so, by distinguishing between time and duration, ultimately fails. Primordial time is not a "qualitative succession," so it is not duration as Bergson defines it. This concept of duration, along with the thesis "time is space," proves that Bergson's attempt to overcome the traditional concept of time takes Aristotle as its point of departure and therefore remains bound to that very concept.

Not only do these discussions of Bergson offer a backdrop for understanding his role in *Being and Time*, but they also offer a sense of the trajectory of Heidegger's own philosophy of time. It is no longer simply a matter, for Heidegger, of distinguishing between the concept of time used by natural scientists and the one employed by historians. The opposition between a homogeneous, quantitative time and a heterogeneous, qualitative time is no longer of fundamental importance. A more crucial distinction is the one between the time in which all natural processes and historical events occur, and the temporality (or "ur-temporality") of Dasein, the kind of being capable of both scientific and historical research. In *Being and Time*, Heidegger will reserve the term "temporality" for the "primordial time" that is the meaning of the being of Dasein, as opposed to the kind of time proper to things encountered in the world. His task will be to exhibit this phenomenon, Dasein's "originary" temporality, as that which grounds its being "ahead of itself," "already in a world," and "together with things"—the structure of care. His task will also be to distinguish originary temporality from time as it is ordinarily understood, and this will require a fuller description of a number interconnected phenomena, from "authentic" temporality to the time "taken care of"; from the "temporalizing" of Dasein to "reckoning with" time; and from "world time" to "the time of nature." One of Heidegger's goals will be to show how, in the ways we go about our daily lives and the ways of thinking we have inherited, we understand being in terms of a particular mode of time: the present.

7. On the Verge of *Being and Time*

We can see now that Heidegger's critical engagement with Bergson begins more than a decade before *Being and Time*, appears in several works concerned directly with time, and becomes more explicit in the courses immediately prior to the writing of *Being and Time*. The engagement begins in 1915 in "The Concept of Time in the Historical Sciences" with Heidegger's use of the distinction between the homogeneous time we measure with clocks and the heterogeneous time we experience, which Bergson calls duration. Heidegger even employs the distinction between quantitative time and qualitative time that he later criticizes as derivative of Aristotle's concept of time. He applies that distinction to the question of how the concept of time functions in the historical sciences as opposed to the natural sciences, and he characterizes historical time, in contrast to the time of nature, in Bergsonian terms: it is qualitative and heterogeneous, as opposed to quantitative and homogeneous.

In 1919–20, in "Comments on Karl Jaspers' *Psychology of Worldviews*," *Phenomenology of Intuition and Expression*, and *Basic Problems of Phenomenology*, Heidegger expresses reservations about Bergson's view of life as something that cannot be grasped conceptually. Heidegger refers disapprovingly to the "Bergsonian" view that reflection on life is bound to distort it, which presupposes that life is unceasing flux and understanding is limited by static concepts that immobilize or de-vivify it. On the contrary, Heidegger believes that it is possible to exhibit the structure of life as a whole by means of a method that is phenomenological and hermeneutical, regarding life as irreducibly meaningful and expressive. Thus, Heidegger's greatest concern, at least at this stage, is not with Bergson's philosophy of time, which he often praises, but with his conception of life as inaccessible and inexpressible.

When Heidegger returns to the question of time in his 1924 lecture "The Concept of Time," he once again employs the Bergsonian distinction between quantitative, homogeneous and qualitative, heterogeneous time. However, he presents quantitative time as dominating not just the natural sciences, but also everyday experience, as illustrated by our use of clocks and preoccupation with how much time we have. While his positive characterization of temporality departs from the conception of time as a flux, his negative account of the concept of time follows Bergson's claim that it is the result of a homogenization of the moments of time, or an assimilation of time to space. The following year, in *History of the Concept of Time*, Heidegger outlines a plan for the destruction of the traditional concept of time, beginning with Bergson's relative transformation of it, and tracing it back through Kant and Newton to its discovery by Aristotle. He links Bergson with Dilthey, Husserl, and Scheler, whose fundamental neglect of the question of being calls for a radicalization of phenomenology. Although Heidegger defers the historical part of his project, he keeps it on the horizon in his *Logic* course of 1925–6 by examining the concept of time in Kant, Hegel, and Bergson. Arguing that all three thinkers depend on Aristotle's account of time in the *Physics*, he finally criticizes Bergson for claiming that time, as we usually understand it, is nothing more than space, and that lived time is a qualitative, rather than quantitative, succession. Here, on the verge of *Being and Time*, Heidegger does not just drag Bergson into the net with other philosophers of life, but uses Bergson's distinction between time and duration as a foil for his own interpretation of temporality.

In these early works, Heidegger acknowledges Bergson's importance for having criticized the traditional concept of time and sought a more original one. Initially, Heidegger follows Bergson's footsteps in distinguishing between quantitative, homogeneous time and qualitative, heterogeneous time—a distinction he later rejects. Taking the full range of his early engagements with Bergson into view, Heidegger displays a deep ambivalence. On the one hand, he praises Bergson's attempt to overcome traditional ways of thinking about time. Heidegger, too, aims to think more radically about time and to show how certain presuppositions about time have shaped the history of philosophy. He even credits Bergson with transforming the concept of time passed down by Aristotle and previously transformed by Newton and Kant. On the other hand, Heidegger portrays Bergson's attempt to overcome the traditional concept of time as a failure, a mere reversal of Aristotle's. He voices concerns about not only Bergson's concept of life, but also his dependence on the categories of quality and quantity, his uncritical acceptance of the notion of time as succession, and his misunderstanding of time as a representation of space. While praising Bergson for having instinctively understood the crucial importance of how we think of time and the need for a radical rethinking of it, Heidegger insists that there is nothing truly original about his concept of duration.

With a better understanding of how Heidegger follows Bergson's footsteps in his early works, we should be in a good position to clarify Bergson's role in *Being and Time*. However, to fully appreciate the impact that Heidegger's early engagement with Bergson had on his thinking about time, we first need to go to the source of Heidegger's interpretation of Bergson, *Time and Free Will*. This is the first work that Heidegger cites when he mentions Bergson in *Being and Time*, and it is the one from which he quotes when he criticizes Bergson in the *Logic* course just prior to it. A closer look at *Time and Free Will* is necessary to determine whether Heidegger is justified in his interpretation of Bergson's distinction between time and duration.

2

Dispelling the Confusion

Pure Duration in *Time and Free Will*

1. Heidegger on the Concept of Duration

Despite a sustained engagement with Bergson's thought lasting through his early Freiburg and Marburg periods, Heidegger dismisses Bergson in *Being and Time* for failing to rethink time radically enough. According to Heidegger, "Bergson's conception [of time] agrees with Hegel's thesis that space 'is' time. Bergson just turns it around: Time (*temps*) is space" (BT 410n/432n). Despite his efforts to overcome the traditional concept of time, Bergson's concept of duration (*durée*) is a mere reversal of Aristotle's concept of time as *arithmos kineseos*, a number by which motion can be measured with respect to the earlier and the later.[1] In this way, Bergson takes Aristotle's account of time and turns it on its head, defining duration in terms of quality rather than quantity. Duration is fundamentally different from number, for Bergson, and what we usually think of as time is really a spatial representation of "pure duration." On Heidegger's interpretation, Bergson fails to get to the root of temporality because, while he calls the traditional concept of time into question, his own concept of duration remains dependent on Aristotle's thinking. This interpretation needs to be examined more critically, in my view, because it is not at all obvious that Bergson's concept of duration is, as Heidegger suggests, a response to Aristotle's account of time.

Heidegger's critique of Bergson deserves closer scrutiny, beginning with the claim that, just as Hegel's interpretation of time in the *Encyclopaedia* follows Aristotle's *Physics*, "Bergson's interpretation of time, too, obviously grew out of Aristotle's treatise on time" (BT 410n/432n). Since the connection between Bergson's thinking and Aristotle's may not be as clear as it is in Hegel's case, Heidegger offers some evidence for this claim. First: "It is not just a matter of an external literary connection that simultaneously with Bergson's *Essai sur les données immédiates de la conscience*, where the problem of *temps* and *durée* is expounded, a treatise of Bergson's appeared with the title: *Quid Aristoteles de loco senserit*" (BT 410n/432–3n). In fact, Bergson did present both his principal thesis (the *Essai*, later translated as *Time and Free Will*) and his Latin thesis (translated as "Aristotle's Concept of Place") to the faculty at the Sorbonne to fulfill the requirements for his doctoral degree in 1889, and both works were published that same year.[2] In "Aristotle's Concept of Place," Bergson offers a close reading of *Physics* IV, 1–9, where Aristotle develops his account of place (*topos*) before moving on to time (*chronos*) in 10–14. Bergson's primary criticism of Aristotle is that "by substituting place for space, he seemingly avoids rather than settles disputes which, in our opinion, refer above all to space."[3] In that essay, Bergson limits his interpretation of Aristotle to the chapters of the *Physics* dealing with place, and he does not discuss time at all. Nor does Bergson make any reference in *Time and Free Will* to Aristotle's account of time, and he mentions Aristotle only once in passing.[4] Consequently, the connection that Heidegger suggests between the two works is far from clear.

Heidegger's second piece of evidence for Bergson's dependence on Aristotle is less circumstantial: "With regard to the Aristotelian definition of time as ἀριθμὸς κινήσεως [*arithmos kineseos*], Bergson analyzes *number* before analyzing time" (BT 410n/433n). Bergson does indeed begin his first detailed discussion of duration in chapter two of *Time and Free Will* with an analysis of the concept of number. His aim is to show that number implies a representation of space, which will allow him to argue that the concept of time as a measurable quantity confuses duration with space. This certainly sets him at odds with Aristotle's concept of time, yet we should not be too quick to accept Heidegger's claim that Bergson conducts his discussion "with regard to the Aristotelian definition of time." For Bergson's primary target is not Aristotle's concept of time, but rather what he calls "time as a homogeneous medium." Bergson's points of reference for this concept of time include Kant's transcendental idealism, empirical psychology, classical physics and astronomy, and Zeno's paradoxes. Given the absence of any discussion of Aristotle in *Time and Free Will*, it is not clear that Bergson's discussion of number is good evidence that he develops his concept of duration with regard to Aristotle.[5]

In Heidegger's view, Aristotle's influence is also evident in two formulas that condense Bergson's thinking. First, Heidegger summarizes the way that Bergson identifies *temps* with space and defines *durée* in opposition it: "Time as space . . . is *quantitative* succession. Duration is described on the basis of a counter-orientation toward *this* concept of time as *qualitative* succession" (BT 410n/433n). Bergson's thinking can thus be reduced to the following formulas: "time is space" and "duration is qualitative succession." These formulas do have a textual basis, for Bergson argues that "time, understood in the sense of a medium in which we make distinctions and count, is nothing but space" (TFW 90/61). He also characterizes duration at one point as "a succession of qualitative changes, which melt into and permeate one another" (TFW 104/69). However, we should note that "qualitative succession" is not a formula for duration that Bergson himself uses. Instead, Bergson prefers the terms "qualitative multiplicity" and "continuous multiplicity" to distinguish duration from any numerical or discrete multiplicity. In an instance more typical of his thinking, Bergson describes duration as "a qualitative multiplicity, with no likeness to number; an organic evolution which is yet not an increasing quantity; a pure heterogeneity within which there are no distinct qualities" (TFW 226/148). Unfortunately, Heidegger says nothing about the crucial concept of multiplicity and the distinction between two kinds of multiplicity at the core of Bergson's account of pure duration in *Time and Free Will*.[6] Nor, even after using the distinction between homogeneous and heterogeneous time in his own early works, does Heidegger address the heterogeneity of pure duration as Bergson conceives it. At best, then, Heidegger's formulas are an oversimplification of Bergson's thinking, and at worst, they are a distortion of it.

As I will show, Heidegger exaggerates the connection between Bergson's thinking and Aristotle's, obscuring much of what is philosophically interesting about the account of duration in *Time and Free Will.* While there is some truth in both of Heidegger's formulas—that "time is space," and duration is "qualitative succession" for Bergson—neither one is faithful to the complexity of Bergson's concept of duration. This is a problem for Heidegger because these formulas are his way of summarizing what makes Bergson's thinking traditional. Intent on distancing himself from Bergson by showing that he falls back into the traditional concept of time, Heidegger says nothing about how he uses duration to uncover the relationship between time, the self, and freedom. In the reading of *Time and Free Will* that follows, I discuss Bergson's distinction between time and duration with regard to its philosophical context and motives. My aim is to show that Heidegger's critique of Bergson, with its narrow focus on how duration is opposed to Aristotle's concept of time, both distorts and oversimplifies Bergson's thinking. Contrary to Heidegger's claims, I argue that Bergsonian duration is no mere reversal of Aristotle's thinking, but rather is the core of

Bergson's strategy for revealing how assumptions about time are involved in some of the perennial problems of philosophy. In *Time and Free Will*, Bergson attempts to shake the foundations of the free will debate by revealing misconceptions about duration and the self that are shared by both sides and to show that the key to resolving that debate is to think about freedom in terms of duration. I also argue that if the concept of time that Bergson rejects can be identified with anyone in particular from the philosophical tradition, it is not Aristotle but Kant. Bergson's aim in criticizing Kant is to challenge not just his account of time, but also his conception of the self. Finally, I argue that Bergson's efforts to rethink both time and the self concretely in terms of duration anticipate Heidegger's own interpretation of temporality in *Being and Time*.

2. Thinking Spatially about Time

In his 1924 lecture "The Concept of Time," Heidegger resolves not to begin with a concept of eternity and consider time in opposition to it, but to "understand time in terms of time." That is, he will approach time not from the standpoint of an extra- or supra-temporal reality, but through the everyday experience of time, especially the kind of experience from which the concept of time originates. This is precisely the strategy that Bergson had already used in *Time and Free Will* to resolve the problem of freedom, which he believes stems, much like Zeno's paradoxes of change and motion, from our habit of thinking spatially about phenomena that are fundamentally temporal.[7] To understand time in terms of time, for Bergson, we need to distinguish it not only from eternity, but also from space. According to him, the debate over free will arises from a false problem based on the common confusion of *duration* with *extension*, of the temporal flow of our experience with the spatial symbols we commonly use to represent it. Bergson suggests that "this confusion once dispelled, we may perhaps witness the disappearance of the objections raised against free will, of the definitions given of it, and, in a certain sense, of the problem of free will itself" (TFW xx/3). His task is to show how this seemingly insoluble metaphysical problem arises from certain intellectual habits, especially the habit of representing time in terms of space.[8]

Thinking spatially about time is, in Bergson's view, one of the most prevalent features of traditional philosophy. In an essay published in 1922, reflecting on the difficulty of understanding time on its own terms, Bergson writes: "All through the history of philosophy time and space have been placed on the same level and treated as things of a kind; the procedure has been to study space, to determine its nature and function, and then to apply to time the conclusions thus reached" (CM 14/5). One of the clearest cases of this, and one to which he returns on many occasions, is Kant's Transcendental Aesthetic, which employs the same set of arguments with minor modifications to show that space and time are pure, *a priori* forms of sensible intuition.[9] Spatializing time is also common in science, especially physics and astronomy, as Bergson understands them. To calculate and predict the movements of bodies, classical physics regards them as moving through time as well as space, a time that can be divided into units and measured. Finally, understanding time in terms of space is a feature of so-called common sense: "duration is always expressed in terms of extension; the terms which designate time are borrowed from the language of space. When we evoke time, it is space which answers our call" (CM 14/5). Because of the predominance of such language and its utility for social life, he contends, it requires a concerted effort to suspend our habit of representing time in terms of space. Bergson intends to demonstrate how such a suspension of spatial thinking is possible and why it is necessary.

Duration is one of the central ideas of Bergson's philosophy, and it precedes his other fundamental concepts of memory and the *élan vital*.[10] As Heidegger rightly recognizes, his distinction between duration and time is meant to clarify the difference between the lived experience of time and conventional representations of it.[11] Duration—alternately described by Bergson as "pure," "true," "real," or "concrete"—is his name for time as it is immediately given to consciousness.[12] In *Time and Free Will*, he writes: "Pure duration is the form which the succession of our conscious states assumes when our ego lets itself *live*" (TFW 100/67); it is "a continuous or qualitative multiplicity with no resemblance to number" (TFW 105/70); it is "a process of organization or interpenetration of conscious states" (TFW 108/72); and it has "no moments which are identical or external to one another, being essentially heterogeneous" (TFW 120/80). According to Bergson, we can *feel* duration immediately through the experience of time passing, but when we measure time, or even divide it into parts, we are no longer dealing with duration but rather space. Duration is, in short, the ceaseless flux of conscious life, of which time, as we commonly understand it, is an abstract representation.

The concept of duration is the primary focus of Bergson's efforts to understand time on its own terms rather than in terms of space.[13] "We usually think in terms of space," he maintains, because "language requires us to establish between our ideas the same sharp and precise distinctions, the same discontinuity, as between material objects" (TFW xix/3). Consciousness involves a steady stream of experiences, which we find it convenient to distinguish and separate from one another as if they had clear boundaries. By thinking of the moments of duration as forming a line or chain reaching from our past into our future, we translate the "confused multiplicity" of inner life into a more manageable form. However, as a consequence of our tendency to think of time as divided into discrete parts, "we find it extraordinarily difficult to think of duration in its original purity" (TFW 106/71). The challenge, for Bergson, is to suspend our habit of representing time in terms of space and recover "the multiplicity of conscious states, regarded in its original purity" (TFW 121/80). Thus, when Bergson speaks of *pure duration*, he means duration that is unmixed with space, or lived time prior to any spatial representation of it. In opposition to a long tradition of philosophers treating time and space as analogous, Bergson tries to get to the root of time—that is, duration—by disentangling it from space.

The concept of time that Bergson attacks is one that arises from thinking about the temporal flux with spatial images and terms. According to him, "when we speak of *time*, we generally think of a homogeneous medium [*milieu*] in which conscious states are ranged alongside one another as in space" (TFW 90/67). This notion of time as a medium in which everything occurs—a quasi-spatial milieu—is familiar and useful. Events "take place" in time, as we commonly say, just as objects are located in space. Anything that happens in time can be "located" with respect to other things that happened before, after, or at the same time. Finite things and processes last for various "lengths" of time—various durations, in the quantitative sense Bergson calls into question—which can be measured just as distance can. However, Bergson thinks that we pay a price for the utility of spatial representations, for they give rise to certain seemingly insoluble problems, including the problem of free will. His aim is to show how the habit of representing time as a "homogeneous medium" like space generates this problem, and then to resolve the problem by dispelling the underlying confusion of time with space.

Bergson's central claim in *Time and Free Will* is that the problem of freedom arises from a triple confusion that is common to those who deny free will and those who defend it. Bergson writes: "[A]ll discussion between the determinists and their adversaries implies a preliminary confusion of duration with extensity, of succession with simultaneity, of quality with quantity" (TFW xix–xx/3). In all three cases, states of consciousness, which occupy time, are regarded as having some of the same characteristics as numerically distinct objects, which occupy space. Bergson argues that the standard ways of both denying free will and defending it rest on the presupposition that time is adequately represented as a homogeneous medium like space. Determinism, as he presents it, entails that every human action can be explained by its causal antecedents, that perfect knowledge of the conditions of an action render that action completely predictable, and that only one course of action is truly possible under a given set of conditions. Determinists represent the decision-making process as a succession of psychological states, the strongest ones strictly necessitating the action to be performed. On the contrary, libertarians claim that various actions are equally possible and that action resulting from a voluntary choice is unpredictable. In Bergson's view, both sides err in assuming that the internal states preceding an action can be adequately represented as discrete terms in a series, and this error stems from the habit of translating temporal processes into spatial terms. Thus, the problem of free will persists because of the prevalence of spatial thinking, and to dispel the triple confusion underlying it, temporality and spatiality must be disentangled.

3. The Confusion of Quality with Quantity and Conscious States with Objects

Although "qualitative succession" is Heidegger's formula for duration rather than Bergson's, it is true that Bergson distinguishes duration from time on the basis of the difference between quality and quantity. In chapter one of *Time and Free Will*, as a prelude to his discussion of duration, Bergson launches an attack on the concept of *intensity* employed by psychologists of his day with regard to inner states. More specifically, he attacks the view that sensations and other states of consciousness possess intensive magnitude. For Bergson, the notion that the intensity of psychological states can be measured rests on a confusion of quality and quantity. Although we commonly think of sensations as increasing or decreasing in intensity, he argues, it is a mistake to suppose that they possess magnitude. He reasons that insofar as a magnitude can increase or decrease in quantity, it must be divisible, and whatever is divisible must be extended. However, states of consciousness have no extension and are neither discrete nor separate from one another.[14] Instead, Bergson maintains, the defining characteristic of conscious states is their interpenetration with other states, which we tend to overlook when we represent them as separate and distinct. His aim is to show that whether we are dealing with deep-seated or superficial feelings, affective or representative sensations, what we interpret as an increase or decrease in *quantity* is actually a change in *quality*. In this way, he sets the stage for his concept of duration as something radically different from time understood as a measurable quantity.

Bergson's discussion of the intensity of conscious states focuses on a crucial difference between lived experience and how we commonly think and talk about it. When we describe a feeling as becoming more intense or growing in intensity, he claims, whether we realize it or not, we have actually experienced a "qualitative progress" involving a multitude of states. For instance, when we experience a joy that becomes more intense, we may tend to think of it as the same feeling growing stronger, but a closer look reveals a transition through a multitude of related feelings. Such joy often begins, Bergson says, "like a turning of our states of consciousness towards the future"; it intensifies as "our perceptions and memories become tinged with an indefinable quality"; and it evolves through "successive stages corresponding to qualitative alterations in the whole of our psychic states" (TFW 10/8). Similarly, sorrow "begins by being nothing more than a facing towards the past, an impoverishment of our sensations and ideas . . . as if the future were in some way stopped up [and] ends with an impression of crushing failure, the effect of which is that we aspire to nothingness" (TFW 11/8). In either case, the experience of a succession of qualitative changes over time comes to be interpreted as a quantitative increase, as if the feeling in question were an extended object taking up a larger portion of psychological space. Not only in "deep-seated" feelings like joy and sorrow, but also in more "superficial" ones like the sensation of effort, Bergson argues, "there is a qualitative progress and an increasing complexity, indistinctly perceived. But consciousness, accustomed to think in terms of space and translate its thoughts into words, will denote the feeling by a single word and will localize the effort at the exact point where it yields a useful result" (TFW 26/19). Thus, at the root of the common notion that sensations increase and decrease in quantity is our habit of spatial thinking.

Bergson attempts to show how the confusion of quality with quantity manifests itself in psychology by offering a critique of psychophysics, a field of research concerned with determining the relationship between physical stimuli and sensations. He regards psychophysics as the epitome of psychology's attempts to explain consciousness through mechanical laws and mathematical formulas, and he aims to show that it rests upon the same confusions that give rise to the problem of free will. Focusing on a series of photometric experiments performed to devise units of measurement for sensations of brightness, Bergson questions whether it is sensations themselves that are measured by the psychophysicists or the physical stimuli that produce them.[15] In his view, their approach suffers from a disregard of our ability to quantify the difference between sensations based on experience with their causes. The psychological states supposedly measured by psychophysicists are not extended objects, Bergson insists, but "we have everything to gain by objectifying these states, by introducing into them, to the largest possible extent, the representation of their external cause" (TFW 70/52). Not only do the psychophysicists take no account of the objectification of conscious states and the introduction of their cause into our estimation of their intensity, but they also push these habits of representation to their theoretical extreme.

Bergson concludes that the concept of intensity actually signifies two different phenomena and that there are two quite different reasons to think of sensations as possessing magnitude. On the one hand, we call an affective sensation more or less intense thanks to a "confused perception," when it is mixed together with a host of other feelings. For instance, a feeling of joy is said to increase when it colors more of our perceptions, thoughts, memories, and other feelings. On the other hand, we call a representative sensation, like the brightness of a light, more or less intense on the basis of an "acquired perception," interpreting qualitative change as quantitative increase or decrease with the help of our experience of cause and effect. For instance, when a light grows brighter, we may think of our sensation of brightness as "growing" because we know that a certain kind of sensation will be produced by a certain magnitude of luminosity. Once again, we interpret the sensation as increasing in intensity, but only because we "transfer the cause to the effect and . . . replace our immediate impressions by what we learn from experience and science" (TFW 54/40). Thus, for Bergson, there are two senses of intensity with respect to psychological states, one referring to a multitude of intermixed feelings or sensations and the other to the magnitude of their external cause.[16] The concept of intensity is therefore a mixture of things that are different in kind, which Bergson envisions as "situated at the junction of two streams, one of which brings us the idea of extensive magnitude from without, while the other brings us from within, in fact from the very depths of consciousness, the image of an inner multiplicity" (TFW 73/54). It is crucial, for Bergson, that we learn to distinguish more carefully between this "inner multiplicity" of conscious states and the kind of multiplicity that is characteristic of external objects.[17]

Bergson's discussion of intensity sets the stage for the arrival of the concept of duration in several ways. First, it establishes the immediacy of qualitative change in consciousness, in contrast to quantitative growth. Prior to any isolation and measurement of conscious states, we experience changes in quality with the passage of time. Intensification is *qualitative progress*, which always involves multiple, interpenetrating states that we nevertheless tend to interpret as a single state increasing in strength.[18] Second, the discussion of intensity calls into question such *habits of representation*, particularly those of objectifying feelings and sensations, and introducing measurable causes or effects into them. Although these habits are useful for understanding and communicating our experience, they obscure the difference between internal and external phenomena, replacing the concrete reality of the former with abstract symbols. Third, Bergson's discussion of intensity leads to the description of consciousness as a *continuous multiplicity*, as opposed to a discrete or numerical one.[19] Although our understanding prefers "clean cut distinctions" and "things with well-defined outlines," he says, "the further we penetrate into the depths of consciousness, the less right we have to treat psychic phenomena as things which are set side by side [*se juxtaposent*]" (TFW 8/6–7). What he intends to problematize is the way we think and speak of things that follow one another in time as if they were juxtaposed in space. This calls for further examination, with the focus shifting from "states of consciousness in isolation from one another" to "their concrete multiplicity, insofar as they unfold themselves in pure duration" (TFW 73/54). For Bergson, the confusion of quality with quantity underlying the notion of intensive magnitude can be traced back to a deeper confusion of duration with extension, which is implicit in the representation of time as an empty, homogenous medium.

4. The Confusion of Duration with Extensity and Time with Space

Heidegger's claim that "time is space" for Bergson must be understood in the light of Bergson's attempt to distinguish duration from our ordinary conception of time. In the second chapter of *Time and Free Will*, Bergson argues that what we usually call "time" is really a confusion of duration with extensity, or a projection of duration into space. His analysis of intensity raised the question of how to conceive consciousness as an "inner multiplicity"—a manifold of sensations, feelings, images, and so forth—without confusing states that "unfold themselves in pure duration" with objects that are arranged in space. According to Bergson, rigorous reflection on immediate experience shows that "the multiplicity of conscious states, regarded in its original purity, is not at all like the discrete multiplicity which goes to form number" (TFW 121/80).

Heidegger offers Bergson's analysis of number in *Time and Free Will* as evidence that his concept of duration is a reversal of Aristotle's concept of time. Bergson does indeed begin his discussion of duration in chapter two with an analysis of number intended to demonstrate that the measurement of time requires that duration be translated into space. Number is, according to Bergson, "a collection of units, or, speaking more exactly, the synthesis of the one and the many" (TFW 75/56). On the one hand, each unit is regarded as provisionally indivisible and identical. On the other, units must be different in order to be added. If they are given simultaneously—here Bergson imagines counting a flock of sheep—they must be distinguished by their location in space, and if not, then by their position in time. In either case, counting involves a juxtaposition of units, and for things to be juxtaposed, or set side by side, they must be separate and distinct from one another. Separation implies that the separate things are extended, so juxtaposition is a spatial relation that obtains only between extended things. Because counting involves juxtaposition, we count by means of implicit or explicit spatial representation. Moreover, number is divisible, and divisibility implies extension as well. Bergson concludes that "every clear idea of number implies a visual image in space" (TFW 79/54), and all counting is done by means of space.

This analysis of number helps Bergson clarify the distinction between two kinds of multiplicity, one pertaining to space and the other to duration. On the one hand, objects occupy space, and as long as they are separated by clear boundaries, "no effort of the inventive faculty or of symbolical representation is necessary in order to count them" (TFW 85/58). On the other hand, psychological states do not occupy space and are not immediately given as separate and distinct, so "we can hardly count them except by some process of symbolical representation" (TFW 86/59). In contrast to a discrete, quantitative multiplicity of objects, inner states thus form a continuous, qualitative multiplicity and must be placed apart or spread out in space in order to be counted. Bergson concludes that while the concept of number is "immediately applicable" to material objects, states of consciousness "cannot be regarded as numerical without the help of some symbolical representation, in which a necessary element is *space*" (TFW 87/59). Just as measuring the intensity of a sensation requires introducing something quantitative into it, such as its cause, enumerating the contents of consciousness requires introducing something spatial. If this is the case, then whenever we count, we either deal with things already situated in space or we engage in some kind of spatial representation.

The difference between these two kinds of multiplicity provides the basis for Bergson's critical analysis of the concept of time. Although we find it convenient to think of our experience as divided into separate and distinct states, he argues, the multiplicity of consciousness is constituted by a continuous stream of experience.[20] Unlike things extended in space, states of consciousness permeate one another and blend together, having no clear boundaries between them. Nevertheless, Bergson puzzles, "when we speak of *time*, we generally think of a homogeneous medium in which our conscious states are ranged alongside one another as in space, so as to form a discrete multiplicity" (TFW 90/61). That is, we tend to represent time as a juxtaposition of moments, even though we experience it as a continuous, unceasing flux. This way of thinking about time, he argues, is common to philosophers, scientists, mathematicians, and ordinary people. Yet there is something problematic, even contradictory, about it: if counting implies space, then "time, understood in the sense of a medium in which we make distinctions and count, is nothing but space" (TFW 91/62). The concept of *time as a homogeneous medium* is thus, for Bergson, a spatial representation of duration.

While the contrast that Bergson draws between duration and number may suggest Aristotle's account of time as the target of his criticism, he is more directly concerned with Kant. In particular, Bergson uses Kant's theory of space in the Transcendental Aesthetic to develop his own conception of space as an empty, homogeneous medium. He notes that since Kant, psychologists have debated whether the qualities of external objects have extension or whether extension is produced by a synthesis of unextended sensations. In his view, both sides presuppose the Kantian distinction between the matter and form of representation, or between sensible qualities, on the one hand, and extension in space, on the other. Bergson argues, under the assumption that sensations are not extended, that our ability to perceive objects as arranged in space depends on a mental act "very much like what Kant calls an *a priori* form of sensibility," an act that "consists essentially in the intuition, or rather the conception, of an empty homogeneous medium" (TFW 94–95/64). As Bergson defines it, "space is what enables us to distinguish a number of identical and simultaneous sensations from one another; it is thus a principle of differentiation other than that of qualitative differentiation, and consequently it is a reality with no quality" (TFW 95/64). The homogeneity of space, at least as humans conceive it, consists in the absence of qualitative differences. While each moment in the life of a conscious being is different, for Bergson, the parts of space are identical in every respect except their location.[21]

Setting Kant aside for the moment, Bergson takes aim at the notion that time, too, is an empty, homogeneous medium. The problem with representing time in this way is that it contradicts our experience of time as continuously passing or ceaselessly flowing. Bergson writes:

> Now, if space is to be defined as the homogeneous, it seems that inversely every homogeneous and unbounded medium will be space. For, homogeneity here consisting in the absence of every quality, it is hard to see how two forms of the homogeneous could be distinguished from one another. Nevertheless, it is generally agreed to regard time as an unbounded medium, different from space but homogeneous like the latter: the homogeneous is thus supposed to take two forms, according as its contents co-exist or follow one another. It is true that, when we make time a homogeneous medium in which conscious states unfold themselves, we take it to be given all at once, which amounts to saying that we abstract it from duration. This simple consideration ought to warn us that we are thus unwittingly falling back upon space, and really giving up time. (TFW 98/66)

Here Bergson accepts the common view that things located in different parts of space exist simultaneously, while things happening at different moments of time occur successively. His concern, however, is that to represent time as homogeneous is to reduce succession to simultaneity. If we admit that states of consciousness follow one another in succession, but that they permeate one another rather than having clear boundaries, then as soon as we think of them as separate and distinct, Bergson claims, we have translated them into a spatial *milieu*. Likewise, when we think of those states as occupying moments that differ only with respect to their position in a series, we have conceded that the parts of time, no less than the parts of space, can be "given all at once." Then what we call "time" is not so much a continuous, ceaseless flux, but a static symbol of it. Consequently, Bergson maintains, "time, conceived under the form of a homogeneous medium, is an illegitimate concept, due to the intrusion of the idea of space upon the field of pure consciousness" (TFW 98/66, translation modified). Our ordinary understanding of time is thus, for Bergson, a translation of pure duration into the language of space.

Because of our habits of representation, Bergson notes, it is extraordinarily difficult to think about time without spatial images or metaphors. For instance, we commonly picture time as a line, extending from the past to the future, with moments separated from one another like geometrical points. Bergson's attack on the concept of time as a homogeneous medium calls into question all such spatial thinking about time. To suspend the habit of representing time in terms of space, a different kind of image is necessary.[22] Bergson suggests that pure duration can be compared more appropriately to a melody, its notes "melting, so to speak, into one another" in rhythmic organization. While the notes follow one another in succession, "we perceive them in one another," and it is their "mutual penetration" that constitutes the melody as such (TFW 100/67). However, as soon as we think of the notes of a melody as separate and distinct, "we project time into space, we express duration in terms of extensity, and succession thus takes the form of a continuous line or chain, the parts of which touch without penetrating one another" (TFW 101/68). The "pure succession" of sounds, blending together in a continuous but heterogeneous flow, comes to be replaced by an *order* of succession. For Bergson, this example illustrates the difference between time as it is lived—pure duration—and time as it is represented in abstract thinking. Any image of the melody as "given all at once" confuses duration with space, in his view, and this applies equally to any temporal process.[23]

To sharpen the contrast between pure duration and time as a homogeneous medium, Bergson offers an analysis of the measurement of time using a clock. This example, like that of the melody, is aimed at showing the difference between the lived experience of time and common representations of it. Interestingly, the focus of Bergson's analysis of the clock as an instrument for measuring time is not where we might expect, namely the dial and the hands, but rather the pendulum. A minute of time, he notes, is measured by sixty oscillations of the pendulum, which can be viewed "all at once by a single mental perception" and represented as "sixty points on a fixed line" (TFW 104/70). However, when a minute is represented in this way, the successive movements of the pendulum are replaced by points given simultaneously in a series, which amounts to juxtaposing the oscillations in an ideal, homogeneous space. If we were to suspend the habit of thinking in terms of space and think in terms of pure duration, Bergson suggests, we would have to refrain from mentally setting the oscillations side by side and regard them instead as flowing into one another. The result would be similar to the experience of hearing a melody, "a continuous or qualitative multiplicity with no resemblance to number" (TFW 105/78). The difference comes into even sharper focus if we contrast the experience of counting sixty seconds by the ticking of a clock with that of being lulled to sleep by it. A feeling of sleepiness cannot be produced by a series of isolated sensations, according to Bergson, but only by the "rhythmic organization of the whole" (TFW 106/79). Each tick of the clock, each movement of the pendulum, alters the *quality* of our conscious experience. Pure duration, as he conceives it, "is not a quantity, and as soon as we try to measure it, we unwittingly replace it by space" (TFW 106/79). Insofar as the function of a clock is to divide duration into units and represent it as a numerical multiplicity, measuring by the clock implies a substitution of homogeneous time for pure duration.[24]

On Bergson's account, the concept of time as a homogeneous medium obscures the fundamental differences between space and duration. Every part of space is present in simultaneity, so there is no trace of the past without a consciousness that *endures* and preserves events that happen in succession. Things that are extended in space possess magnitude, but they lack the rhythmic organization of the moments of duration. Objects can be divided without altering their nature, whereas psychological states are essentially altered by isolation. In summary, objects in space are characterized by "mutual externality without succession" and conscious states, which endure, by "succession without mutual externality" (TFW 108/81). Despite these differences, Bergson acknowledges, we tend to regard objects as enduring and states of consciousness as extended. His explanation is that a process of "endosmosis" occurs by which we project our duration into space, on the one hand, and the extensity of objects invades our consciousness, on the other. Bergson concludes that this exchange between mind and world "gives rise to a symbolic representation of duration, derived from space" (TFW 110/82).[25]

Bergson proceeds to argue that the confusion of duration with extensity and time with space is evident not only in common sense, but in ancient philosophy, classical physics, and mathematics. He is especially concerned with the traditional understanding of motion as homogeneous and divisible, like space rather than duration. Bergson claims that Zeno's paradoxes involving motion arise from the failure to distinguish between a movement—a simple, indivisible act—and the space it traverses. In the paradox of Achilles and the tortoise, it seems as if the runner cannot overtake his slower competitor, and according to Bergson this is because Achilles's movement is identified with homogeneous space, which is infinitely divisible. The paradox can be resolved if we recognize that Achilles does not have to perform a series of discontinuous advances from one position in space to the next to pass the tortoise; he only has to perform the kind of quick, fluid motion of which he is capable. A real movement is neither discontinuous nor divided, Bergson argues, but is a continuous change. Only when a movement is represented in terms of space, he claims, can it be "divided and recomposed according to any law whatever" (TFW 113/84). In reality, concrete movement, like pure duration, is a continuous, qualitative multiplicity that can be abstractly symbolized as a series of positions in space. With such symbols, we interpret quality as quantity and project heterogeneous duration into homogeneous space. According to Bergson, the puzzle of how Achilles can overtake the tortoise arises from the translation of movement, which occurs in duration, into an infinitely divisible line or segment of space.[26]

Bergson also detects the replacement of duration with a spatialized concept of time in the way that physicists and astronomers measure velocity by counting "simultaneities" (TFW 114–17/76–79). First, they mark the beginning and end of a movement as two distinct points of simultaneity between the moving thing and the observer. Then they measure the distance between the positions occupied at these moments, ignoring the "interval of duration" experienced by the conscious observer. Here again, a continuous movement and the duration in which it occurs are translated into a stretch of space traversed by the moving thing in abstract, homogeneous time. The disregard for concrete duration in this kind of operation is clear from the fact that it would make no difference if all motion were accelerated uniformly: "Consciousness would have an indefinable and as it were qualitative impression of the change, but the change would not make itself felt outside consciousness, since the same number of simultaneities would go on taking place in space" (TFW 116/87).[27] This leads Bergson to conclude that physics treats time and motion in such a way that it eliminates duration and mobility, dealing only with space and simultaneities.

Finally, Bergson sees something similar in mathematics, illustrated by algebra in particular. Just as physicists bracket the consciousness of duration and treat motion as something occurring in the medium of abstract time, mathematicians must isolate, within the flux of experience, relations and operations that always obtain. As such, mathematics is concerned with whatever is stable across various instances of motion and intervals of duration. For example, even though "it is of the very essence of duration and motion, as they appear to our consciousness, to be something that is unceasingly being done," on the contrary, "an algebraic equation always expresses something already done" (TFW 119/79). That is, like physics, mathematics leaves out the interval of lived time, or its passage, in order to deal with something fixed and permanent. Even calculus, which is supposed to deal precisely with change, concerns itself with points separated by smaller (infinitesimal) intervals, according to Bergson. It still has to translate real change, motion, and duration into a spatial or homogeneous medium.

Bergson offers these analyses to show that philosophy and science tend to overlook the difference between the two forms of multiplicity, qualitative and quantitative, and the two conceptions of time they entail. The distinction can be summed up in these terms: on the side of immediate experience is the continuous and heterogeneous multiplicity of pure duration; on the side of thought mediated by symbols is the discrete and homogeneous multiplicity characteristic of number, space, and abstract time. The traditional concept of time deserves to be called "abstract" because it arrests the continuous flow of inner life and reinscribes it in a series of discrete points. According to Bergson, just as the confusion of these two multiplicities is responsible for the Eleatic philosophers' view that time and motion are illusory, it is also at the root of the problem of freedom. To resolve Zeno's paradox, we need to recognize that *motion* can only be divided and recomposed arbitrarily through the substitution of *an immobile medium*, namely the space traversed. Similarly, to resolve the problem of free will, we need to recognize that *action* can only be analyzed by replacing it, as well as the agent, with *abstract, inert symbols*. For Bergson, the temporal processes of action and decision making need to be distinguished from certain spatial representations of them with which they are commonly confused.

Before considering Bergson's attempt to demonstrate how the confusion of duration with extensity underlies the problem of free will, let us return for a moment to Heidegger's critique of Bergson in *Being and Time*. Heidegger says that Bergson defines duration as "qualitative succession," reversing Aristotle's concept of time by replacing the category of quantity with its opposite. This, along with Bergson's analysis of number, is supposed to prove that his concept of duration is directed against Aristotle's definition of time. Heidegger is correct that Bergson uses the categories of quality and quantity to distinguish duration from time, and he is also correct that Bergson's analysis of number places him in opposition to the notion that time is a measurable quantity. However, it is not at all clear that Bergson's purpose for analyzing number is to attack Aristotle's account of time as *arithmos kineseos*. His explicit aim is to demonstrate that all counting presupposes spatial representation and that the measurement of time treats it as a *homogeneous medium* like space. In this regard, Bergson is more directly concerned with Kant's treatment of time and space than Aristotle's, as will become even clearer when we examine Bergson's critique of Kant in the conclusion of *Time and Free Will*.

Heidegger focuses on Bergson's definition of duration as "qualitative succession," but this is an oversimplification for two reasons. First, the distinction between number as a quantitative multiplicity and duration as a qualitative one is only one dimension of Bergson's account. Even after using the distinction between homogeneous and heterogeneous time himself in both "The Concept of Time in the Science of History" (1915) and "The

Concept of Time" (1924), Heidegger says nothing about the *heterogeneity* of duration for Bergson. His interpretation therefore fails to take seriously the extent to which Bergson, by characterizing duration as "pure heterogeneity," thinks about time in terms of identity and difference.[28] Nor does Heidegger consider Bergson's arguments about the *continuity* and *indivisibility* of duration, as opposed to the infinite divisibility of space and the discontinuity of things juxtaposed in it. These features of duration are equally crucial, for Bergson, because they characterize "inner life" prior to any representation of it in terms of space. Second, Heidegger's formula is an oversimplification because while Bergson uses the term "succession" with regard to duration, he does not conceive it as a "succession of nows," but is in fact sharply critical of linear representations of time. His intention in identifying duration with succession is to indicate a kind of "pure succession," which is continuous change or movement, with our states of consciousness permeating one another rather than being juxtaposed like a series of discontinuous points. Because it is crucial for Bergson that moments of duration not be confused with points on a line, he uses the term "qualitative multiplicity" to mark the difference between lived time and time as we commonly represent it.

Heidegger also says that "time is space" for Bergson, a particularly puzzling claim, given how much effort Bergson devotes to differentiating between lived time, or duration, and homogeneous space. This formula also oversimplifies Bergson's thinking and even distorts it by failing to acknowledge *what kind of time* is at issue. It is impossible to glean from Heidegger's critique in *Being and Time* what Bergson really means when he says that time "is nothing but space." In the first place, Bergson's concern is with a *common conception* of time, which he regards as an abstract, spatial representation of concrete, lived time. Heidegger is aware of this, for when he first mentions Bergson in the introduction to *Being and Time*, he pledges to restore the common conception of time (*vulgäre Zeitbegriff*) to "its rightful autonomy—contrary to Bergson's thesis that time understood in the common way is really space" (BT 17/18).[29] But Heidegger's condensation of Bergson's view to "time is space" completely obscures the fact that Bergson is dealing with two conceptions of time, one at the level of ordinary language and thought and the other at the level of immediate experience. As Heidegger surely understands, Bergson's point is not simply that time *is space*, but that time *is commonly represented in terms of space*, or that what we usually call "time" is a projection of duration into space. His concern is to show how duration in its "original purity" gets covered up by spatial thinking and replaced by a symbolic representation—in other words, how the pure heterogeneity of our experience gets translated into a homogeneous medium, a spatial *milieu*. In short, Bergson's claim is not that "time is space," but that the concept of *time as a homogeneous medium*, implicit in much of our thinking, *is a spatial representation*. His efforts to distinguish between time and pure duration are therefore aimed precisely at disentangling temporality and spatiality.

5. The Fundamental Self and the Superficial Self

The philosophical importance of the distinction between pure duration and time as a homogeneous medium as it is developed in *Time and Free Will* lies in the way Bergson employs it to think about selfhood and freedom. The problem of free will is rooted, he claims, in our failure to distinguish between "two forms of multiplicity, two very different ways of regarding duration, [and] two aspects of conscious life" (TFW 128/95). In his view, the confusion of duration with space that we find in the concept of time as a homogeneous medium is also evident in common representations of the self or ego (*le moi*). According to Bergson, there is the self as it unfolds in pure duration, on the one hand, and the self whose conscious states are separate and distinct, perhaps even measurable by psychologists, on the other. As such, "conscious life displays two aspects according as we perceive it directly or by refraction through space" (TFW 137/91): first, the "deep" or "fundamental self" (*moi fundamental*), whose states "dissolve into and permeate one another without any precise outlines" (TFW 132/87), and second, the "superficial self" (*moi superficiel*), "a self whose existence is made up of distinct moments, whose states are separated from one another and easily expressed in words" (TFW 138/91–92). While the latter is "much better adapted to the requirements of social life in general and language in particular" (TFW 128/85), the concept of duration makes it possible to see the former by opposing the tendency of most of our language and thought to divide conscious life into impersonal states. If duration is a "counter-concept," as Heidegger suggests, this reveals what Bergson develops it in opposition to: not only the traditional concept of time, but also a conception of selfhood that it supports. Bergson argues that if we can think about conscious states in terms of pure duration rather than as forming a discrete, numerical multiplicity, then it is possible "to recover this fundamental self" (TFW 129/85).[30]

Bergson has argued that consciousness consists of a continuous or qualitative multiplicity, unfolding in duration. In contrast, space is a homogeneous medium in which we discover a discrete or numerical multiplicity of extended objects. His analysis of duration and motion leads him to conclude "that space alone is homogeneous, that objects in space form a discrete multiplicity, and that every discrete multiplicity is got by a process of unfolding in space," and finally, "that there is neither duration nor succession in space" (TFW 120/80). If we accept that every discrete multiplicity implies space, then it follows that when we regard our experience as divided into conscious "states" or "contents," this already projects consciousness into space. In fact, Bergson argues, so much of the language we use to talk about conscious life implies divisibility that we can hardly avoid the confusion that he aims to dispel.[31] Moreover, if things in space exist simultaneously and not successively, then our view of objects likewise "borrows" something from our duration. Thanks to the double movement of endosmosis, Bergson contends, not only do we think of our own inner life as a collection of separate states, but we also think of external objects as enduring. On the one hand, we impose the form of spatial extension on consciousness when we divide it into discrete states, represented as a numerical multiplicity. On the other, we impose the form of temporal continuity—in which the past is preserved and permeates the present—on the external world when we think of objects as enduring. Thus, endosmosis involves both the reification of conscious life and the representation of things as unfolding in time or, in other words, participating in our duration.

If catching a glimpse of pure duration demands that we suspend the habit of representing time in terms of space, then this is also what is required, on Bergson's account, for recovering the fundamental self. So far, he has focused on showing how homogeneous time is a symbolic representation of duration, which translates the pure heterogeneity and rhythmic organization of conscious life into spatial language and images. The next step is to demonstrate how our confusion of duration with space influences our common conception of the self or ego. According to Bergson, endosmosis between the "succession without mutual externality" of consciousness and the "mutual externality without succession" of space influences our representations of the self just as much as those of time:

> In a word, our ego [*moi*] comes in contact with the external world at its surface; our successive sensations, although melting into one another, retain something of the mutual externality that belongs to their objective causes; and this is why our superficial psychological life is unrolled in a homogeneous medium without costing us a great effort. But the symbolic character of this rep-

> resentation becomes more striking as we penetrate further into the depths of consciousness: *the internal ego, which feels and is impassioned, which deliberates and decides, is a force whose states and modifications permeate one another and undergo a profound alteration as soon as we separate them from one another to unroll them in space.* . . . Thus the mutual externality which material objects gain from their juxtaposition in homogeneous space reverberates and is propagated into the depths of consciousness: little by little our sensations are detached from one another like the external causes that gave birth to them, and our feelings or ideas like the sensations with which they are contemporaneous. (TFW 125–6/83–84, emphasis added)

Here Bergson distinguishes between two aspects of the self, the fundamental and the superficial, corresponding to the distinctions between qualitative and quantitative multiplicity, and between pure duration and time as a homogeneous medium. If thinking in terms of space distorts our understanding of time, he suggests, it equally distorts our understanding of the self. The self that "unrolls" in space is only one aspect of conscious life, the shallowest, most impersonal one.

What exactly is this "fundamental" or "deep" self? First, it is a phenomenon that is usually hidden from view: "Below homogeneous duration [i.e., time] . . . a duration whose heterogeneous moments permeate one another; below the numerical multiplicity of conscious states, a qualitative multiplicity; below the self with well-defined states, a self in which *succeeding each other* means *melting into one another* and forming an organic whole" (TFW 128/85). It is not a separate thing but an *aspect* of conscious life, because it "forms one and the same person with the superficial ego" (TFW 125/83). We get an indication of what it is like from our dreams, in which the connection between the mind and the external world is relaxed, submerging measured time in the feeling of duration and the distinct multiplicity of objects in a confused multiplicity of images and impressions.[32] The fundamental self appears prior to any division of consciousness into states and mediation by general terms. Each of its feelings is something completely singular "which lives and develops and is therefore constantly changing" (TFW 133/88). For the fundamental self, every sensation, feeling, thought, and act is continuous with every other experience and yet qualitatively different from it. It is a whole that differs at every moment by virtue of pure duration, a self that is ceaselessly growing, changing, and becoming something new.

If the fundamental self lives in pure duration, then the ego as it is usually represented is, Bergson says, "the shadow of the self projected into homogeneous space" (TFW 128/85). This superficial self is one "whose existence is made up of distinct moments, whose states are separated from one another and easily expressed in words" (TFW 138/91–92). It is a spatial representation of the self, one that we mistake for the real thing for the simple reason that it is better suited to the needs of communication and social life. While it serves many practical purposes to render singular feelings into general terms, Bergson maintains, ordinary language does not convey the nuances of inner life. Rather, language makes our feelings and sensations communicable by abstracting from what is most personal about them and attending to that which "is public property and thereby belongs to space" (TFW 162/107). Thus, the superficial self consists of a discrete multiplicity of impersonal elements designated by generic terms, an "outer crust of clean-cut psychic states, which are separated from one another and consequently fixed" (TFW 167/110). As with the concept of time as a homogeneous medium, this aspect of the self is a symbolic representation of a reality that is heterogeneous and in constant flux. The problem arises, according to Bergson, when the representation is mistaken for reality, the abstract for the concrete, and the dynamic and a static thing for the dynamic and progressive unfolding of life. He claims that this is what happens with certain presuppositions maintained by both parties to the free will debate, namely the associationist theory of mind and the conceptions of action and decision following from it. In Bergson's view, the self is conceived by both determinists and libertarians as "refracted through space," when instead it should be understood in terms of duration.

Bergson's distinction between two aspects of the self, corresponding to pure duration and time as a homogenous medium, is crucial to his project in *Time and Free Will*, yet remarkably Heidegger says nothing about it. Consequently, his critique of Bergson's concept of duration ignores its philosophical motives and context. This concept, as Bergson originally develops it, serves the purpose of showing that the free will debate arises from traditional assumptions about the self involving time. According to him, the relationship between the self and its acts cannot be understood adequately on the basis of representations of the self that spatialize or reify it. What makes Heidegger's silence on Bergson's recovery of the fundamental self especially problematic is that it conceals the way that Bergson anticipates Heidegger's interpretation of the self in *Being and Time*. As Bergson does with the fundamental self, Heidegger presents authentic "being-oneself" as a phenomenon hidden by public interpretation. The counter-phenomenon,

Heidegger's inauthentic "they-self," is analogous to Bergson's spatialized self, both being reflections of the things we encounter in the world that dominate common sense and philosophical thinking.

6. Freedom: Getting Back into Duration

To see more clearly how Bergson's account of the two aspects of the self anticipates Heidegger's treatment of the self in *Being and Time*, we need to examine his interpretation of freedom. Bergson's strategy for dealing with the problem of free will is to reveal certain assumptions concerning time and the self from which it arises. His first target is the conception of the self presupposed by determinists, a self whose conscious states follow one another *mechanically* according to causal laws. After criticizing the associationist theory of mind supporting determinism, Bergson briefly outlines his view of freedom. He then argues, with regard to three questions concerning free will, that libertarians also engage in a kind of spatial thinking that eliminates duration and favors determinism. Because both sides of the debate wrongly assume that time can be adequately represented by space, they are compelled either to deny freedom or to define it improperly. Bergson proposes to settle the debate by rethinking freedom on the basis of the fundamental self and its activity in concrete duration. Free acts are as much a feature of our experience as the passage of time, according to Bergson, yet as expressions of the whole person, they occur rarely and are not absolute.

Bergson begins his analysis of the problem of free will by arguing that determinism is based on a mechanistic theory not only of the universe, but also of the self. He first distinguishes between *physical* determinism, the view that freedom is "incompatible with the fundamental properties of matter, and in particular with the principle of the conservation of energy," and *psychological* determinism, the view that "our actions are necessitated by our feelings, our ideas, and the whole preceding series of our conscious states" (TFW 142/94). Both kinds of determinism assume that the principle of conservation applies universally, to consciousness as well as matter, and this ignores the difference between duration and homogeneous time. He argues that the principle "can only be intelligibly applied to a system of which the points, after moving, can return to their former positions" (TFW 152/101), yet this appears not to be the case for consciousness. For going back to a previous state—in other words, reversibility—is possible in space but not duration. Every conscious state "is reinforced and swollen by the whole of its past"; therefore, Bergson concludes, "[w]hile past time is neither a gain nor a loss for a system assumed to be conservative, it may be a gain for the living being, and it is indisputably one for the conscious being" (TFW 153/101–2).[33] To assume that the principle of conservation applies to consciousness is thus to replace duration with abstract, homogeneous time.

Bergson argues that reification of the self is a common thread between determinism and associationism. Having distinguished between physical and psychological determinism, he claims that the former is reducible to the latter because it maintains that the principle of conservation applies to consciousness, that "the determination of conscious states by one another is absolute" (TFW 148/99). This "psychological hypothesis" at the basis of all determinism finds support in the associationist theory of mind, according to which every succession of ideas is governed by certain laws of association. On this model, the self is "a collection of psychic states, the strongest of which exerts a prevailing influence," with all such states "presented as distinct things which there is no inconvenience in naming separately" (TFW 159/105). Associationists thus commit the error, Bergson argues, of treating sensations, feelings, and ideas as separate and distinct, which means representing them as juxtaposed in space. This theory of mind replaces the "concrete phenomenon" of consciousness with an "artificial reconstruction" (TFW 163/108) and covers up the fundamental self with "a phantom self, the shadow of the ego projecting itself into space" (TFW 165/111). By abstracting conscious life from duration, associationism provides determinism with a "mechanical conception of the self" (TFW 171/113) whose actions follow necessarily from antecedent states.

Bergson's idea of freedom depends on the rejection of this mechanistic view of the self. According to psychological determinism, every action is determined by the strongest idea or feeling influencing an agent at the moment of decision. Bergson counters that although this may be true in most cases in our daily lives, certain actions are the manifestation of an inner state that reflects the "whole soul" (TFW 165/109). On his account, an action is more or less free depending on whether it issues from the "surface" of consciousness or its "depths." Many impressions do produce ideas and determine actions without disturbing anything beneath the surface, and one responds to them like a "conscious automaton" (TFW 168/111). But when an action "agrees with the whole of our most intimate feelings, thoughts, and aspirations, with that particular conception of life which is the equivalent of all our past experience" (TFW 170/112), then we call it free. For Bergson, determinists mistake the superficial, "outer crust" of the self for the fundamental self, but libertarians tend to err in the opposite extreme if they maintain that all our actions are free. According to Bergson, few of our actions spring from the fundamental self, so freedom is rare, and it is by no means absolute. Automatism and self-determination differ in degree, not in kind. Bergson's radical proposition is that the root of the mistakes on both sides of the debate is "a certain misconception of duration" (TFW 173/114). He defends this by showing how three versions of the free will debate pose the question in terms that replace duration with abstract, homogeneous time.

One version of the free will debate involves a disagreement over the contingency of action. According to libertarians, whenever we act freely, it is the case that we could have acted otherwise; that is, other actions were "equally possible." Determinists claim, on the contrary, that "given certain antecedents, only one resultant action was possible" (TFW 174/115). According to Bergson, this formulation of the question implies a spatial representation of the decision-making process leading to an action. In the "clumsy symbolism" of common sense, we think of an agent facing a decision as standing before two contrary actions, X and Y, and we represent the deliberation prior to a choice as a line splitting in two different directions at the moment of decision. The question of whether the agent, having chosen X, could have chosen Y instead, is a badly posed one, Bergson argues, because "[t]o ask such a question is to admit the possibility of adequately representing time by space and a succession by a simultaneity" (TFW 180/119). In the first place, the very distinction between two actions, X and Y, is artificial, as is the representation of them as lying in two opposite directions, or positions in space. The actions are thus reified, "transformed into indifferent inert things awaiting our choice" (TFW 177/116). Next, the depiction of the agent's hesitation, deliberation, and choice as a forking path is a spatialization of "the living activity of the self." When we represent the series of events leading to an action in this way, Bergson claims, "we perceive succession under the form of simultaneity, we project time into space, and we base our reasoning consciously or unconsciously, on this geometrical figure. But this figure represents a *thing* and not a *progress*; it corresponds, in its inertness, to a kind of stereotyped memory of the whole process of deliberation and the final decision arrived at" (TFW 180–1/119). Finally, a figure of this sort regards the decision-making process in retrospect, so it "does not symbolize the time which is passing but the time which has passed" (TFW 182/119–20). As in Zeno's paradoxes, duration is replaced with abstract time, and the process is depicted as entirely in the past, as if the future were already given. Unrealized possibilities are thereby conceived as preexisting realities, implying that the possible is prior to the real.[34] In this version of the free will debate, determinists have the upper hand because their opponents join them in supposing that nothing is lost in the translation of concrete duration into abstract time.

Another version of the free will debate revolves around the predictability of action. Determinists commonly argue that if we knew all the conditions affecting an agent at the moment of decision, we could predict a future action with certainty. In contrast, libertarians define a free act as "that which could not be foreseen, even when all the conditions were known in advance" (TFW 220/144). Bergson attempts to show that this formulation is equally problematic for the reason that it eliminates duration and, like the previous question, assumes that time can be adequately represented in terms of space. According to determinists, a person's actions are perfectly predictable with the benefit of complete knowledge of the antecedent conditions of an action, so any lack of predictability is due to incomplete knowledge. But what is entailed in complete knowledge of *everything* leading up to a person's action? Bergson argues that, to know everything influencing one's choice, we would need to know all the psychological conditions of her action, including the intensity of every feeling involved. Because intensity is a matter of quality rather than quantity, to know the intensity of another person's feelings, we would have to experience them in their duration. Bergson concludes that complete knowledge of the conditions of a person's action would be possible only for an observer who had the same personal history, that is, an observer who was identical with the agent. The argument from predictability, however, rests on the assumption that a person is a collection of impersonal states, the strongest ones resulting in action, and that their intensity can be known by an external observer. This presupposes a view *sub specie aeternitatis*, from which all the moments of time can be surveyed simultaneously and a person's future is "a road already marked out across the plain, which we can contemplate from the top of the mountain without having traversed it" (TFW 191/125).[35] The question of predictability thus involves three illusions according to Bergson: that the intensity of psychological states is a calculable quantity, that a dynamic process is no different from a series of mechanical acts, and that time can be adequately represented by space.

Bergson acknowledges that there is something "natural and almost unavoidable" in these illusions thanks to the predictive successes of science. If astronomical calculations make it possible to foresee events thousands of years in the future, then why should we doubt the predictability of a person's actions in the near future? According to Bergson, the analogy between human action and astronomical events is weak because it ignores the crucial difference between abstract time and the concrete duration of conscious life.

For the prediction of astronomical events, scientists use equations that eliminate the immediate experience of duration. If time were accelerated, Bergson claims, it would alter our conscious experience but not our astronomical predictions. The reason is that "the symbol *t* does not stand for a duration, but for a relation between two durations, for a certain number of units of time, in short, for a certain number of *simultaneities*" (TFW 193/127). The same simultaneities, or relationships between coexisting bodies, would occur, but the intervals between them would be shorter. In a sense, what astronomers do when they predict events in the distant future is theoretically accelerate or leap over great intervals of time. To predict a future event is thus to eliminate the wait—the experience of duration—between then and now, and to see it as it will appear to hypothetical future observers. But for consciousness, duration always makes a difference, and it cannot be eliminated without producing a qualitatively different experience.[36] A feeling of suspense that lasts for a few minutes can be enjoyable, for example, whereas one that lasted for days might seem unbearable. While we may think of these as different *quantities* of the same feeling, the difference in *quality* is what we experience in pure duration. Our so-called states of consciousness are dynamic, "alive and therefore constantly changing," says Bergson, and for that reason "it is impossible to cut off a moment from them without making them poorer by the loss of some impression, and thus altering their quality" (TFW 196/129). It is also impossible to know the precise result of a feeling, in the way that astronomers can predict the trajectory of a body moving through space, without having endured it oneself. In short, when we are dealing with conscious agents and dynamic processes, we cannot dispose of time as we please. The problem with the question of predictability is this: "when we ask whether a future action could have been foreseen, we unwittingly identify that time with which we have to do in the exact sciences, and which is reducible to a number, with real duration, whose so-called quantity is really a quality, and which we cannot curtail an instant without altering the nature of the facts which fill it" (TFW 197–8/130). Because the prediction of future events on the basis of present conditions assumes that duration can be replaced by homogeneous time, it is another badly posed question.

The third and final version of the free will debate that Bergson tackles is concerned with causality. Determinists maintain that, regardless of whether it can be foreseen, a future act must be determined by antecedent conditions, for everything is subject to the law of causality, and the same causes always produce the same effects. The question is whether the "same causes" ever recur in consciousness. Bergson allows that a set of physical conditions can be reproduced identically, as in a laboratory experiment, but he insists that psychological states cannot. Because of the radical heterogeneity of duration, "we cannot here speak of identical conditions because the same moment does not occur twice" (TFW 200/131). We may call two feelings by the same name, but in terms of duration, "the same feeling, by the mere fact of being repeated, is a new feeling" (TFW 200/150). For this reason, the concept of causality that we use to understand the physical world may not be applicable to consciousness. As Bergson has already argued, the notion that the principle of conservation governs conscious states just as it does material objects is problematic because it assumes a permanence and impermeability that they do not possess. Similarly, he suggests, "if the causal relation still holds good in the realm of inner states, it cannot resemble causality in nature" (200–1/132).

On Bergson's analysis, the assumption that a series of psychological states (in duration) is organized in fundamentally the same way as a series of physical states (in space) is based on the confusion of two senses of causality. The first sense, derived from observation of the external world, indicates a relationship between things that appear in regular succession. One phenomenon is said to be the cause of another if, whenever the former occurs, the latter always follows. The second sense of causality is derived from reflection rather than observation, and it indicates a relationship between ideas rather than objects. One phenomenon is said to be the cause of another in this sense if the former contains the latter, or the latter is implied in it. Causality in the first sense is an objective but merely probable connection between physical phenomena, Bergson explains, while causality in the second sense is a certain but merely subjective connection between inner states. However, when we think of causality as a connection that is both objective and necessary, we "pass imperceptibly from the first meaning to the second, and we picture the causal relation as a kind of prefiguring [*préformation*] of the future phenomenon in its present conditions" (TFW 204/134).

In modern philosophy, the conflation of these two senses of causality results in two competing notions of how the future is "preformed" in the past. On the one hand, future events are thought to be "virtually contained" in external phenomena in the way a geometrical figure's properties are contained in its definition. Bergson argues that this mathematical conception of causality borrows its necessity from the relation of identity, which is apparent in Spinoza's idea of eternity. For an intelligent observer who can survey the world *sub specie aeternitatis*, the future is bound to the present (A causes B) with the same necessity that the present is bound to itself (A = A). This conception of causality as *necessary connection* reveals a tendency in philosophy "to transform relations of succession into relations of inherence, to do away with active duration, and to substitute for apparent causality a fundamental identity" (TFW 208–9/137). The attempt to impose the necessity of logical and mathematical relations on nature is, in temporal terms, an attempt to make the future present and to eliminate duration. On the other hand, the future is thought to be preformed in the present in the way that an action, flowing continuously from an idea, is already contained in that idea, as "an effort which has not yet issued in an act" (TFW 214/161). This model of prefiguring is not mathematical but psychological, beginning with the idea of effort; and rather than a mechanistic view, it is a dynamic conception of causality as *force*. Leibniz represents the culmination of this view, regarding even objects as souls and their qualities as perceptions or states of consciousness, with the future being given in the present through the preestablished harmony among them. Here again, the future is given and duration is eliminated. Bergson argues that what he has described as the endosmosis between mind and world is also evident in the way that the idea of force, originating in our feeling of inner effort, is projected outward onto external objects, and the idea of necessary connection, originating in mathematical relations, is introduced into the succession of our inner states, strengthening the mechanical conception of the self.

Ultimately, for Bergson, all three versions of the free will debate presuppose that *time can be adequately represented in terms of space*. This can be seen in the way freedom is implicitly defined in the three badly posed questions. To define a free act as one that "might have been left undone" is to imply that possibilities preexist their realization and to maintain "an absolute equivalence between concrete duration and its spatial symbol" (TFW 220/144). To define a free act as one that "could not be foreseen, even when all the conditions were known in advance" (TFW 220/144) is to assume that one could have perfect knowledge of another's inner life from an external standpoint. This presupposes that conscious decisions are as regular and predictable as planetary orbits, that psychology is not essentially different from astronomy, and that duration can be replaced by homogeneous time. Finally, to define a free act as one that "is not necessarily determined by its cause" (TFW 220/144) is to assume that the same causes can reappear in consciousness and that two moments can be identical, like points in geometrical space. Moreover, the concept of a cause is a mixture of two very different elements, the confusion of which reinforces the conception of the self as a collection of states in which the future is already given, thus a superficial, spatial distortion of the fundamental self that experiences itself acting freely in the face of an open future.

If we deny that time can be adequately represented by space, reject badly posed questions, and think in terms of duration, what does it mean for an act to be free? Bergson does not offer a definition of freedom, but his attempt to reveal the misconceptions about duration involved in the free will debate offers several clues for understanding freedom in temporal terms. First, free acts are performed by the fundamental self, arising from the whole soul, which means that they are the fruit of our entire past, as opposed to automatic or mechanical reactions to the present situation. Second, because so much of our behavior *is* reactive, bypassing the depths of the self, free acts are extremely rare. Third, Bergson does not set up a simple dichotomy between freedom and necessity, voluntary and involuntary action, as if a different self were responsible for each. Rather, he claims that reflexive behavior is "the substratum of our free activity" (TFW 168/111) just as a body's organic functions provide the basis for conscious life. Free acts are therefore never absolutely free, but there are degrees of freedom. And since it is impossible to distinguish sharply between the heterogeneous elements of duration, the boundary between free choice and automatism is never so clear that it cannot be called into question. Fourth, because free acts are the culmination of a "natural evolution" (TFW 171/113) in conscious life, they do not appear suddenly or *ex nihilo*, but gradually through dynamic processes, and they change the self that performs them. As such, the fundamental self may endure, but it is nothing static or permanent.[37]

Two analogies help to convey Bergson's interpretation of freedom as the creative evolution of an individual. On the one hand, freedom is like growth and maturation, so that free acts "drop" from a person "like an overripe fruit" (TFW 176/116). Just as the fruit cannot be completely explained by the stage prior to it—the flower—in which it is only vaguely prefigured, free acts cannot be explained completely on the basis of antecedent states.[38] On the other hand, freedom is like artistic expression in the sense that free acts have an "indefinable resemblance" to the person who does them, and they bear the "mark of our personality" (TFW 172–3/129–30) just as a work of art reveals the style of the artist. Like artworks, free acts cannot be reconstructed from a collection of parts or phases—or, for that matter, reasons or motives—because it is only in retrospect that they can be regarded as contained in their antecedents. Freedom is something immediately given in our experience, as long as we do not abstract that experience from the flow of time. But our habit of representing the series of states preceding an action with spatial images, as if the process were a thing whose parts are all given at once, confuses succession with simultaneity, eliminates duration, and obscures freedom.

What is radical about Bergson's thinking on the issue of freedom is the strategy by which he counters the notion that free will is an illusion. His approach is not simply to defend a conception of the will as a faculty of choice undetermined by any causal antecedents. It is to show that the very terms in which the debate is conducted rest on certain misconceptions about temporality. The root of these misconceptions is the assumption that time can be adequately represented in terms of space or that duration can be eliminated without a loss. It is on the basis of this assumption that both sides of the free will debate think spatially about action and decision, and ultimately about the self that acts and decides. Although they defend contrary positions with respect to free will, determinists and libertarians share the same metaphysical prejudice that time is not fundamentally different from space. On Bergson's analysis, freedom is not an illusion, although it is commonly misunderstood; the illusion is the self whose future is already given in the present. Pure duration, as opposed to time as a homogeneous medium, implies that the future is open. While this openness has been regarded by many as a subjective appearance, the fault of a temporal perspective, Bergson maintains that it is irreducible.

7. Reversing Kantianism

Bergson concludes *Time and Free Will* with a brief but forceful critique of Kant, which makes it even more difficult to accept Heidegger's claims about his dependence on Aristotle. Bergson objects that even though Kant is no determinist, his defense of freedom removes it from time and the self that endures. On Bergson's interpretation, Kant secures the status of the world of appearances as a lawful order by defining causality as an *a priori* condition of experience and therefore a necessary connection between phenomena. Kant maintains that causality and freedom are not incompatible, but that freedom must be a property of things in themselves—specifically, the soul—while causality is a connection between appearances.[39] Thus, Kant resolves the problem of free will, in Bergson's view, by transporting freedom beyond space and time, for the sake of preserving the universal validity of causality in nature and safeguarding science. In this way, Kant guarantees knowledge by restricting it to the spatio-temporal world, where everything is determined by natural laws, including the empirical ego and its actions. Bergson's concern is that Kant's solution to the problem of free will is "to put freedom outside of time and to raise an impassable barrier between the world of phenomena, which he hands over entirely to our understanding, and the world of things in themselves, which he forbids us to enter" (TFW 235/153). In other words, Kant defends freedom, but in a way that makes it inaccessible and incomprehensible.

What leads Kant to conceive freedom as beyond time, according to Bergson, is that he understands time in terms of space: "Kant's great mistake was to take time as a homogeneous medium . . . the very distinction which he makes between time and space amounts at bottom to confusing time and space, and the symbolical representation of the ego with the ego itself" (TFW 232/151). In this respect, Kant is no different from determinists who assume that time can be adequately represented in terms of space. For one thing, Kant presents time as the form of inner sense, represented as an empty, infinite continuum in which phenomena are juxtaposed like objects in space. While this conception of time makes the phenomenal self knowable, it supports the notion that "the same states can recur in the depths of consciousness, just as the same physical phenomena are repeated in space" (TFW 232/151), and that conscious life must be governed by causality just as the external world is. For Bergson, Kant overlooks the fact that science treats time as a measurable quantity only by translating all duration into simultaneities. Although Kant distinguishes time from space by its irreversibility, nothing in his account of time would prevent it from being radically accelerated for the purposes of calculation and prediction. Thus, Kant follows common sense and science in conceiving time as a homogeneous medium, failing to distinguish it from pure duration and substituting spatial representations of time for our lived experience of it.

The result of Kant's failure to distinguish duration from space, Bergson argues, is that despite his attempt to save free will, he ends up characterizing the self that can be known—the empirical ego—as a conscious automaton. This conception of the self, which Bergson attributes to the associationists as well, follows directly from the spatialization of time: "In place of a heterogeneous duration whose moments permeate one another, we thus get a homogeneous time whose moments are aligned in space. In place of an inner life whose successive phases, each unique of its kind, cannot be expressed in the fixed terms of language, we get a self which can be artificially reconstructed" (TFW 237/154). For Bergson, Kant concedes too much by conceiving of the self as divided into discrete states or representations that are causally connected. Presupposing, as determinists do, that time is a homogenous medium and that it is adequately represented with spatial images, Kant differs only in the respect that instead of denying free will altogether, he transports it beyond time.

As an alternative to determinism, on the one hand, and Kantianism, on the other, Bergson challenges some traditional assumptions about freedom and attempts to rethink it by thinking about time and the self concretely in terms of duration. This requires distinguishing lived time from any spatial representation of it and understanding it in terms of pure duration: "A qualitative multiplicity, with no likeness to number; an organic development which is not yet an increasing quantity; a pure heterogeneity within which there are no distinct qualities" (TFW 226/148). In duration, moments succeed one another without being external to one another, and each moment is open to all the others. This conception of time in its "original purity" points us in the direction of a non-mechanical conception of the self: "a living self, whose states, at once undistinguished and unstable, cannot be separated without changing their nature, and cannot receive a fixed form or be expressed in words without becoming public property" (TFW 236/154). The recovery of this self from abstract, spatial thinking is a task for philosophy, in Bergson's view, but it is equally a challenge for living. Not only do we misunderstand freedom when we fall prey to thinking of ourselves and our time in terms of space, but we also forget what it means to act freely.

Understanding time in terms of duration rather than space thus turns out to be more than a strategy for resolving a metaphysical problem. For Bergson, thinking in terms of duration is a matter of existential significance. Thanks to the habit of representing our sensations and feelings, thus the moments of our lives, as if they were juxtaposed in space, he says, "we are rarely free. The majority of the time we live outside ourselves, hardly perceiving anything of ourselves but our own ghost, a colorless shadow which pure duration projects into homogeneous space" (TFW 231/151). The more we objectify, solidify, and externalize our inner states, we more inclined we are to think of them as mechanically determined, to behave reactively, and to forfeit our freedom. In Bergson's view, Kant enables this by understanding time in terms of space and transporting freedom to "the atemporal domain of 'things in themselves'" (TFW 238/155). But freedom cannot be outside of time; on the contrary, for Bergson, "[t]o act freely is to recover possession of oneself, to replace oneself in pure duration" (TFW 231–2/151).[40] To experience freedom, it is not necessary to transcend the phenomenal world, but instead "to carry ourselves back in thought to those moments of our lives when we made some serious decision, moments unique of their kind, which will never be repeated" (TFW 238–9/155). These moments of truth, which stand out in our memory, are nothing like points on a line, for they enable us to regard the present as a future that might not have been.

Although Bergson does not introduce the project of *Time and Free Will* as a confrontation with Kantianism, his concluding discussion of Kant both clarifies his analysis of how the confusion surrounding time engenders metaphysical problems and anticipates his later efforts to rethink the relationship between mind and body in terms of memory. In an attempt to provide an alternative to determinism and libertarianism as well as to Kantian and associationist psychology, Bergson proposes a kind of reversal of Kant's Transcendental Aesthetic. If, as Kant contends, our understanding of the world is organized by the *a priori* forms of time and space, then it is probable, Bergson claims, that our self-understanding is mediated by "certain forms borrowed from the external world" (TFW 223/146). In other words, if spatio-temporality is a framework provided for matter by the mind, then it stands to reason that matter leaves its mark on the mind in a reciprocal fashion; once again, this is the exchange that Bergson calls *endosmosis*. Accordingly, each of the main concepts analyzed in *Time and Free Will*—intensity, time (duration), and causality (determination)—involves a "compromise" between internal and external phenomena. First, it is the compromise between qualitative change (a feature of psychological states) and quantitative difference (a feature of external objects) that results in the confused mixture known as "intensive magnitude." Next, it is the compromise between duration (mind) and extensity (matter) that "introduces succession into external things" and "externalize[s] the successive moments of our inner duration in relation to one another" (TFW 228/149). This compromise enables us to experience objects as enduring, on the one hand, and states of consciousness as separate and distinct, on the other, with both occurring in homogeneous time. In a sense, science has to break this compromise and dissociate external phenomena from duration to make them available for mathematical calculation. Bergson argues that metaphysics, too, should break this compromise by dissociating inner life from space and restoring the traits of duration to consciousness. If Kant's great achievement was to recognize how the subject conditions objects and the external world, Bergson challenges us to recognize the intrusion of external forms on our understanding of time, selfhood, and freedom. This reversal of Kant's thinking, which is Bergson's answer to the transcendental turn, is far more interesting and important than the reversal of Aristotle of which Heidegger accuses him.

8. Anticipating Originary Temporality

By now I hope to have shown that what Heidegger says is obvious about Bergson's concept of duration, namely that it is a mere reversal of Aristotle's concept of time, is anything but obvious. There may be a significant link between *Time and Free Will* and "Aristotle's Concept of Place," but the absence of any discussion of Aristotle in the former makes it difficult to determine what the nature of that link is. Bergson's analysis of number is also not good evidence that Aristotle is the target of his criticism. That analysis is intended to show that counting presupposes spatial representation, and the measurement of time treats it as a homogeneous medium like space. Bergson's target is the habit of representing time in terms of space, a feature of both common sense and philosophical thinking, which appears as early as Zeno's paradoxes and as recently as the associationist theory of mind. It is not at all obvious to what extent Bergson views Aristotle's account of time as a manifestation of the intellectual habit responsible for the problem of freedom.

What *is* obvious is that Kant plays a more prominent role than Aristotle in *Time and Free Will*. Whereas Bergson mentions Aristotle only once in passing, he discusses Kant at length and even presents his own distinction between duration and time against the backdrop of Kant's Transcendental Aesthetic. Bergson accepts Kant's view of space, which he interprets as "the intuition, or rather the conception, of an empty homogenous medium" (TFW 64/94–95), but he rejects the notion that time differs from space only in the respect that it has one dimension rather than three. Kant is mistaken, according to Bergson, to treat both time and space as homogeneous, for "we cannot finally admit two forms of the homogeneous, time and space, without first seeking whether one of them cannot be reduced to the other (TFW 66/99). In Bergson's view, Kant's account of time confuses duration, which is irreducibly heterogeneous, with space. Bergson's critique of Kant in *Time and Free Will* suggests that if his concept of duration is directed against any thinker's concept of time, it is not Aristotle's but Kant's. Now, for Heidegger, it should be possible to trace Bergson's thought back to Aristotle's by means of his response to Kant, insofar as "the Kantian interpretation [of time] moves within the structures developed by Aristotle" (BT 25/26). This would be a way to overcome the problems I have pointed out with establishing the connection between Bergson's concept of duration and Aristotle's concept of time. However, the fact remains that Heidegger asserts a direct link between Bergson's thought and Aristotle's on the basis of insufficient evidence, rather than demonstrating how Bergson's critique of Kant displays an indirect link to Aristotle.

With regard to the two theses that Heidegger attributes to Bergson, I have not shown that they are false, but rather that they need to be amended. To borrow an expression from Heidegger, it is correct but not true that Bergson characterizes duration as "qualitative succession"—that is, it is not faithful to the complexity and originality of his thinking. In Bergson's own formulations, duration is *continuous* or *qualitative multiplicity*, *pure heterogeneity* and *rhythmic organization*, its moments *permeating* one another. The qualitative-quantitative distinction is indeed one aspect of Bergson's account, but it is only a point of departure. Furthermore, when Bergson identifies duration with succession, he distinguishes between the lived experience of succession, which is continuous change or movement, and the representation of succession as a line or series of discontinuous points. Heidegger's formula is misleading insofar as it glosses over this distinction and makes it seem as if Bergson subscribes to the very image of time he rejects. There is something even more deeply misleading about Heidegger's claim that "time is space" for Bergson. It would be more accurate to say that on Bergson's account, *time as a homogeneous medium is a spatial representation*, or that *time is commonly represented in terms of space*. Indeed, the extent to which Heidegger's formula distorts Bergson's view is clear from the way Bergson, during his discussion of the problem of free will, insists that *time cannot be adequately represented by space*. As such, Bergson develops the concept of duration precisely to clarify the fundamental difference between concrete, lived time and homogeneous space.

Not only does Heidegger's critique of Bergson in *Being and Time* radically oversimplify his thinking, but Heidegger also pays no attention to the ways in which Bergson's project in *Time and Free Will* anticipates his own interpretation of temporality. He says nothing about Bergson's description of duration as time in its "original purity," an expression that evokes Heidegger's own search for "originary (*ursprünglich*) temporality." He says nothing about Bergson's phenomenological descriptions of the measurement of time and his arguments for why the time that is measured is not "pure duration." He says nothing about Bergson's distinction between two aspects of the self, one that meets the requirements of social life and another more fundamental one that is covered up by it. Finally, Heidegger says nothing about Bergson's strategy of resolving the problem of free will by revealing the presuppositions about time from which it arises and rethinking freedom in terms of time—or more precisely, duration—rather than space. While it may be somewhat unfair to fault a thinker for omissions made for the sake of brevity, these omissions are troubling because they have helped to ensure the success of Heidegger's attempts to distance himself from Bergson.

Time and Free Will is by no means Bergson's final word on duration, but it does clearly display his motives for developing the concept. His concern is not just that time has been misunderstood, but that the confusion of time with space is responsible for some of the basic problems of philosophy, particularly the problem of free will. According to him, those who defend freedom and those who deny it share a mechanical conception of the self that rests on the assumption that time can be adequately represented by space. On both sides, the self that deliberates, decides, and acts is viewed as analogous to an object moving through space. Thus, for Bergson, the representation of the self as a thing whose changes are causally determined arises from the representation of time as a homogeneous medium like space. Contrary to what we commonly assume, he argues, there is a fundamental difference between something essentially temporal, like the self, and something spatial, like the images often used to depict conscious decision. Bergson develops the concept of duration to clarify the differences between time as it is immediately given, or lived, and time as it is commonly represented. In his view, the problem of free will can only be resolved if we see through the thing-like self whose states are divided in abstract time (thus space) to the living, evolving, enduring self.

While there are many genuine differences between Heidegger and Bergson, we are now in a good position to draw attention to some significant similarities. With the concept of duration, Bergson tries to uncover and illuminate the origin of what we usually call "time." He calls into question both how time has been conceived by philosophers and how it is usually understood by ordinary people. As such, Heidegger's project in *Being and Time*, division two, of showing that "[the traditional] concept of time and the common understanding of time in general originate from temporality" (BT 17/18) aims at succeeding where he thinks Bergson fails in *Time and Free Will.* There, Heidegger follows Bergson not only in distinguishing between an "originary" and a "derived" time, but also in seeking to demonstrate how assumptions about time lie at the root of some of the most basic problems of philosophy. Under Bergson's influence—although certainly not his influence alone—Heidegger comes to view time as not just one philosophical issue among others, but a central issue for ontology. Like Bergson, he believes that thinking through the primordial phenomenon of time is a necessary step toward the reorientation of philosophy, a field dominated by the assumption that being means presence.

3

Uncovering the Primordial Phenomenon

Originary Temporality in *Being and Time*

1. Bergson in *Being and Time*

Early in *Being and Time*, Heidegger declares his opposition to Bergson's view that "time understood in the common way is really space" (BT 17/18). Heidegger also frames his project of uncovering originary temporality, the primordial phenomenon of time, as an attempt to rethink "the traditional concept of time that has persisted from Aristotle to Bergson and beyond" (BT 17/18, translation modified). Yet it is difficult to see the extent to which Heidegger's interpretation of temporality in *Being and Time* is shaped by his engagement with Bergson for several reasons. For one thing, there is no decisive confrontation with Bergson, but rather a handful of scattered references and a sketch of a critique in a pregnant footnote near the end. With the exception of that footnote, Heidegger's remarks about Bergson are brief and relatively obscure. Another reason is that Heidegger's main concern seems to be to distance his interpretation of temporality from Bergson's theory of duration. Like Bergson, he intends to radically rethink time, and he does not want their views to be confused. Consequently, Heidegger offers a less ambivalent but also less nuanced take on Bergson in *Being and Time* than he does in the lecture courses surrounding it. Yet another reason is that although he lodges several complaints against Bergson in division one, the crux of his critique does not become clear until his discussion of the genesis of the ordinary concept of time near the end of division two. Even then, he explains how Aristotle's concept of time "levels off" and "covers up" the primordial phenomenon of time but says nothing explicit about Bergsonian duration. Finally, Bergson's role in *Being and Time* is less than obvious because Heidegger also engages with a number of other thinkers to whom he is more willing to acknowledge his debt. As such, the influence of Husserl, Dilthey, Kierkegaard, Hegel, Kant, Augustine, and Aristotle on Heidegger makes it more difficult to identify Bergson's contributions.[1] The complexity of Heidegger's interpretation of temporality, involving a network of interconnected phenomena, makes the task of understanding Bergson's role in *Being and Time* even more formidable.[2]

Despite these difficulties, the prominence of Heidegger's remarks about Bergson early in *Being and Time* and his promise of a more decisive confrontation in the second part suggest that Heidegger's project is in some ways a response to Bergson's efforts to understand time in its original purity. Having appropriated key insights of Bergson's in "The Concept of Time" and acknowledged his importance as a pivotal thinker in *History of the Concept of Time*, at the very least Heidegger regards Bergson as a cautionary model. That is, Heidegger aims to avoid Bergson's mistake of attempting to get to the root of time by merely reversing the traditional concept of time. Yet the very strategy of posing philosophical problems—in Heidegger's case the "question of being"—in terms of time displays Bergson's influence. Like Bergson, Heidegger aims to reveal not just a misunderstanding about the nature of time, but also deep assumptions about time shared by the entire philosophical tradition. Whereas Bergson had sought to dispel the confusion between duration and extension, Heidegger seeks to uncover "originary temporality." As such, while there is little overt resemblance between temporality as Heidegger understands it and Bergsonian duration, Heidegger still needs to warn us against confusing them. Ironically, Bergson is only a cautionary model for Heidegger to the extent that his thinking is also exemplary, at least in some respects. My aim is not to show that Bergson was the sole or even the most important influence on Heidegger's account of temporality in *Being and Time*, but to clarify Heidegger's concerns about Bergson and to reveal how, even as Heidegger's path leads in a different direction, he continues to think about time with and against Bergson. From the introduction to *Being and Time*, where Heidegger casts Bergson as an heir to the Aristotelian concept of time, to the final chapter, where he subordinates Bergson's thinking to Hegel's, he is intent on distancing his interpretation of temporality from Bergson's theory of duration. In so doing, however, Heidegger draws attention to the proximity between them. Heidegger intends to show how the temporality of Dasein lies at the origin of our common understanding of time, just as pure duration does for Bergson. However, instead of beginning with an attack on the concept of time, Heidegger develops an existential-ontological account of temporality and then shows how the traditional concept of time is derived from it. He argues that the primordial phenomenon of temporality is covered up by the concept of time as an infinite, irreversible succession of nows. This temporality is implicit in our understanding of time but hidden by our interpretation of time as something objectively present, like a substance or thing.

My reading of *Being and Time* begins with what Heidegger explicitly says about Bergson, usually in the form of terse proclamations about what he misunderstands. I argue that Heidegger's dismissal of Bergson rests on three major complaints already expressed in his earlier works: Bergson's neglect of ontology, his treatment of the concept of time as illegitimate, and his dependence on traditional, Aristotelian categories. After clarifying Heidegger's critique, I show how Heidegger's efforts to think more radically about time lead him to characterize originary temporality as a double movement reflected in the structure of care. I also examine his distinction between authentic and inauthentic temporality, which sets the stage for his account of the derivation of time from temporality. To clarify Bergson's role in *Being and Time*, I focus on Heidegger's discussion of time in the final chapter of division two. There, Heidegger finally examines the difference between our ordinary understanding of time and the temporality of Dasein, and he locates duration at the level of what he calls "world time." I argue that the deepest concern underlying all of Heidegger's complaints about Bergson, and thus the crux of his critique, is this: Bergson believes that time is misunderstood because it is confused with space, but the issue is really the privilege of presence. Nevertheless, I maintain, even as Heidegger contrasts Bergson's thinking with his own, he is also concerned with disentangling temporality from spatiality. Despite the serious concerns he expresses about Bergson's attempt to rethink time, Heidegger employs a strategy similar to Bergson's for thinking about selfhood in terms of time, which is apparent in the way he characterizes authenticity and inauthenticity. Although Bergson appears at first glance to be a minor character, I hope to show that he plays a leading role in Heidegger's efforts to put time at the center of his ontology.

2. Time and the Question of Being

Heidegger's account of temporality in *Being and Time* circumvents many well-known questions about time raised by philosophers. What is time? Is time absolute or relative? Is time real, or is it an illusion? Did time have a beginning, and will it come to an end? Heidegger avoids these questions not because they are uninteresting, but because his approach to time is dictated by what he considers to be a more fundamental question. Heidegger's project in *Being and Time* is to reawaken the question of *the meaning of being in general*, or of what unifies the many senses in which something may be said to be, and to formulate this question concretely. His first task is "the interpretation of *time* as the possible horizon for any understanding whatsoever of being" (BT xxix/1). It is no longer a matter, as it was at an earlier stage for Heidegger, of contrasting the concepts of time used in the natural and the human sciences. Now his ambitious goal is to demonstrate "that *time* is that from which Dasein tacitly understands and interprets something like being at all . . . the horizon of every understanding and interpretation of being" (BT 17/17), and even "*that—and in what way—the central range of problems of all ontology is rooted in the phenomenon of time correctly viewed and correctly explained*" (BT 18/18). Time is central to Heidegger's project of fundamental ontology because, as he intends to show, being has always been interpreted by philosophers with regard to time, but in a way that has escaped notice. Questions such as "what is time?" and "is time absolute?" presuppose that time *is* something to begin with. Instead of assuming that time is a particular kind of being or a property of beings, Heidegger calls into question the relationship between being and time.[3]

In the introduction to *Being and Time*, Heidegger revises the plan announced in *History of the Concept of Time* in a way that excludes Bergson. In part one of *Being and Time*, he proposes to pursue the question of the meaning of being by examining three things: first, the basic ontological structures of Dasein, the kind of being that understands being (division one); second, the temporal meaning of these structures, or their implicit temporality (division two); and third, the temporality of being as such (division three, which was not published). He also presents a new plan for a "phenomenological destruction of the history of ontology" (BT 37/39) in part two (also unpublished), still dealing with the concept of time in Kant and Aristotle but no longer beginning with Bergson. Heidegger's aim for the destruction is to show "that the ancient interpretation of the being of beings is oriented toward the 'world' or 'nature' in the broadest sense, and that it indeed gains its understanding of being from 'time' " (BT 24/25). More specifically, he will argue that ever since the ancient Greeks, "[b]eings are grasped in their being as 'presence' ['*Anwesenheit*']; that is to say they are understood with regard to a definite mode of time—the '*present*' ['*Gegenwart*']" (BT 24/25). This shows up in the Greek terms that Aristotle uses for being in the primary sense: *ousia*—translated into Latin by medieval scholars as *substantia*—and *parousia*—translated as *praesens*. There is also a temporal meaning implicit in the terms that Parmenides and Plato use for how beings are intellectually grasped: *legein* and *noein*, discourse and intuition, which both refer to "the simple apprehension of something objectively present in its sheer objective presence [*Vorhandenheit*]" (BT 24/25). Through the modern period and beyond, Heidegger claims, *knowledge* is interpreted as a kind of *making present* [*Gegenwärtigen*] and *substance* as permanence or *constant presence* [*ständige Anwesenheit*].[4]

Heidegger thus follows Bergson in maintaining that time needs to be rethought, but his concern is to show how the structure of temporality has been obscured by the traditional interpretation of being as presence, or the assumption that *being equals being present.* Before he focuses directly on time, his existential analysis of Dasein both reveals and resists the tendency to think of ourselves, the world, and others—even of time itself—as "present-at-hand" or "objectively present" (*vorhanden*). By approaching Dasein as a kind of being that is neither a mere object nor an isolated subject, but "being-in-the-world," Heidegger challenges the assumption that being equals being objectively present. His interpretation of Dasein's temporality then situates the mode of "making present" within an "ecstatic openness" that he views as the origin of time. He argues that the traditional concept of time as a succession of "nows" is formulated when "time itself is taken to be one being among others" thanks to "an understanding of being which is oriented toward time in an inexplicit and naïve way" (BT 25/26).[5] It will only become clear how time is the horizon for our understanding of being once temporality has been uncovered and freed from the privilege of presence that defines the philosophical tradition going back to the ancient Greeks. Heidegger thus attempts to think more radically about time by distinguishing it from originary temporality, a primordial phenomenon that both shows through and is covered up by the concept of time. In other words, he attempts to do what he claims Bergson fails to do: uncover the origin of time without falling into traditional ontological assumptions that have obscured it from the view of philosophers.

Just as Heidegger does not deal directly with many of the perennial questions about time, he deals with Bergson's philosophy indirectly for the most part. His approach is to distance himself by linking Bergson with other philosophers of more immediate concern to him. While it would have been possible, and in some ways much easier, to show how Bergson's interpretation of time emerges from his disagreements with Kant and other modern thinkers, Heidegger continues to associate Bergson with Aristotle. He makes this connection in the introduction to *Being and Time* (§5) along with his first reference to temporality:

> We need an original explication of time as the horizon of the understanding of being, in terms of temporality as the being of

> Dasein which understands being. This task as a whole requires that the concept of time thus gained be distinguished from the common understanding of it. The latter has become explicit in an interpretation of time which reflects the traditional concept that has persisted since Aristotle and beyond Bergson. We must thereby make clear that, and in what way, this concept of time and the common understanding of time in general originate from temporality. In this way the common concept of time receives again its rightful autonomy—contrary to Bergson's thesis that time understood in the common way is really space. (BT 17/17–18)

This is the first indication in *Being and Time* of the fundamental distinction between time (*Zeit*) and temporality (*Zeitlichkeit*). Heidegger ultimately argues that time, understood as an infinite, irreversible succession of "nows," is a concept that "springs from" (*springt aus*) the temporality of Dasein—thus, "primordial" or "originary temporality" (*ursprünglich Zeitlichkeit*). We can see another crucial distinction here between the "traditional concept of time" (*traditionellen Zeitbegriff*) and the "ordinary" or "common understanding of time" (*vulgäre Zeitverständnis*) from which it arises.[6] According to Heidegger, the traditional concept provides a more explicit, thematic formulation of our ordinary understanding of time as something that we have or don't have enough of, that we refer to with every "now" and "then," and that we measure with clocks and other instruments.[7]

In his discussion of the history of ontology (§6), Heidegger credits Aristotle with being the first thinker to formulate the traditional concept of time in the *Physics*. He also calls Aristotle's philosophy the "scientifically highest and purest stage" (BT 25/26) of ancient Greek ontology. While the interpretation of being as presence appears earlier, notably in Parmenides, it is Aristotle who designates *ousia* (substance) as the primary sense of being. Likewise, Aristotle provides a definition of *chronos* (time) and an analysis of the link between *chronos* and *kinesis* (motion) that sets the agenda for future philosophical thinking about time. According to Heidegger:

> Aristotle's treatise on time is the first detailed interpretation of this phenomenon that has come down to us. It essentially determined all the subsequent interpretations of time, including that of Bergson. (BT 25/26)

Heidegger offers Kant's interpretation of time as a leading example and goes on to outline the critical destruction of the ontological tradition planned for

Being and Time, part two.[8] As already mentioned, this tradition is defined for Heidegger not only by a concept of time, but also by the *privilege of presence*. No one, according to Heidegger—not Hegel, not Bergson, not even Husserl—has succeeded in rethinking time radically enough to have overcome the problems with time first raised by Aristotle.

After the brief but pregnant remarks linking Bergson to Aristotle in the introduction to *Being and Time*, Heidegger waits until the final chapter of division two before offering any clarification of their connection. His discussion of Aristotle's concept of time in §81 does not explain how Bergson's thinking about time is "essentially determined" by Aristotle's. In the penultimate section of division two (§82), Heidegger chooses to focus on Hegel's thought rather than Bergson's, effectively replacing Bergson with Hegel as the most important recent heir to Aristotle's concept of time.[9] In contrast to his plan in *History of the Concept of Time* to take Bergson as the point of departure because of his "relative transformation" of this concept, here he designates Hegel's account "the most radical way in which the vulgar concept of time has been given form conceptually" (BT 407/428). However, during his discussion of time and spirit for Hegel, in a footnote in which he argues that "Hegel's concept of time is even drawn *directly* from Aristotle's *Physics*" (BT 410n/432n), Heidegger digresses long enough to dismiss Bergson. This footnote contains the most substantial discussion of Bergson in *Being and Time*, which is still only a rough sketch of a critique and a condensation of arguments presented in *Logic: The Question Concerning Truth*.[10] As in that course, Heidegger subordinates Bergson's thought to Hegel's and claims, "Bergson's interpretation of time, too, obviously grew out of an interpretation of Aristotle's treatise on time" (BT 410n/432n). In the discussion that follows, Heidegger points to Bergson's thesis that time, as it is commonly understood, is space, and his definition of duration as qualitative succession as evidence that Bergson merely reverses Aristotle's understanding of time. I have attempted to show that the connection that Heidegger alleges between Bergson and Aristotle with respect to time is not so obvious. In any case, Heidegger's point is that Bergson's attempt to rethink time is not as radical, original, or innovative as it might seem.

Heidegger concludes his discussion of Bergson: "To the extent that anything essential has been gained at all beyond Aristotle and Kant, the concern is more with grasping time and 'time consciousness'" (BT 410n/433n). As the terms "grasping time" (*Zeiterfassung*) and "time consciousness" (*Zeitbewußtsein*) are ones employed by Husserl in his 1904–5 lectures on the phenomenology of time, this remark appears to connect Bergson with Husserl as well. We have seen how Heidegger already suggests such a connection in *History of the Concept of Time* (SS 1925),[11] where he criticizes Dilthey, Husserl, and Scheler for a "fundamental neglect" of the question of being. Time-consciousness becomes an issue briefly in *Logic: The Question of Truth* (WS 1925–6), when he says:

> Some philosophers, in the course of expounding the phenomenon of time, especially recently, have made a distinction between objective and subjective time, or between transcendent and immanent time. This distinction says nothing, however, because it could well be that the so-called "immanent" time, the time of lived experience, merely carries over into itself the characteristics of the so-called transcendent or objective time" (LQT 204–5/246).[12]

While Husserl is the most likely target of this objection, Bergson may also be in Heidegger's sights, considering his characterization of duration as "lived time," an "inner multiplicity" of conscious states. In any case, Heidegger's claim in *Being and Time* that Bergson deals primarily with the *consciousness* of time implies that his approach suffers from the same problems as Husserl's, but Heidegger unfortunately does not elaborate.

In the first chapter of division one of *Being and Time* (§10), Heidegger mentions Bergson during a discussion of *Lebensphilosophie* and personalism that condenses critiques of Dilthey, Husserl, and Scheler worked out in *History of the Concept of Time*.[13] Heidegger notes that modern philosophy is often said to begin with Descartes and the "*cogito sum*," the subject whose knowledge rests on a foundation of self-certainty expressed in the motto "I think, therefore I am." Descartes deals with the *cogito* and its *cogitationes*, Heidegger argues, without adequately addressing "the ontological question of the being of the *sum*" (BT 45/45), a question that his successors also fail to raise. Instead, even the most radical recent and contemporary thinkers approach the "I am" as an ego or subject whose basic ontological character is unquestioned. Even philosophers who criticize Descartes's conception of subjectivity by rejecting "the 'substantial soul' or the 'reification of consciousness"—including Dilthey, Husserl, Scheler, *and* Bergson—offer no positive account of "the nonreified *being* of the subject, the soul, consciousness, the spirit, the person" (BT 46/46). For Heidegger, it is therefore necessary to avoid using all these terms, as well as 'life' and 'human being,' and to focus on the basic character of Dasein, the kind of being to which they all refer.[14]

According to Heidegger, the *Lebensphilosophie* and personalism movements are not exceptions to the rule that the being of Dasein is implicitly presupposed, not ontologically clarified, in both philosophical and scientific investigations of life. Dilthey deserves recognition for making life the focus of his philosophical efforts, refusing to reduce it to "psychic elements and atoms," and instead seeking "to understand its 'experiences' in their structural and developmental interconnections" (BT 46/46). Scheler resists the reification of consciousness by making his theme the person, which for him "can never be thought of as a thing or a substance" (BT 47/47), but must be treated as a performer of acts. However, neither thinker gives sufficient attention to the question of the *being of life* or the *being of the person*, just as Husserl never raises the question of the *being of consciousness* or intentionality. In Heidegger's view, these thinkers are confined by the limits of a tradition "thoroughly colored by the anthropology of Christianity and the ancient world, whose inadequate ontological foundations personalism and the philosophy of life also ignore" (BT 47/73/48). Within this tradition, long before Descartes's inauguration of modern thought, the being that says "I am" is understood as human: as a rational animal (ξώον λόγον έχον [*zoon logon echon*]), a finite being (*ens finitum*) and creature of God (*ens creatum*). Even the most radical attempts to distinguish persons from things remain limited by what is considered self-evident in this conception of the human being. Thus, Heidegger claims, "along with Dilthey and Bergson, all the directions of 'personalism' and all tendencies toward a philosophical anthropology influenced by them share these limits" (BT 46/72–73/47). As sympathetic as he is to these philosophers, he contends that their thinking is limited in the same way as the sciences of anthropology, psychology, and biology are, namely by their failure to answer, or even pose, "the question of the *kind of being* of this being that we ourselves are" (BT 49/50).

Here, the ambivalence that Heidegger displays toward Bergson prior to *Being and Time* seems to have given way to a one-sided concern with distancing himself from Bergson. Heidegger is anxious to distinguish his project from biologically, anthropologically, and psychologically oriented philosophies that rest on "inadequate ontological foundations." Bergson's focus on life, an obvious similarity with Dilthey, and his influence on Scheler's concept of the person make it convenient to charge him with the deficiencies in their approaches. In contrast to *History of the Philosophy of Time*, Heidegger gives no sense in *Being and Time* of Bergson's having positively influenced phenomenology, only of his having neglected the question of being as Dilthey, Scheler, and Husserl do. Likewise, Heidegger makes no mention of the importance of Bergson's attempt to overcome the traditional concept of time, only his dependence on Aristotle. Thus, it now appears that Heidegger's goal with respect to Bergson is to establish his guilt by association—first with the Western philosophical tradition as a whole and its dominant conception of time, and second with the contemporary movements of *Lebensphilosophie* and personalism—of a fundamental neglect of ontological issues, and most crucially the question of the meaning of being.

The cost of neglecting the question of being, Heidegger suggests, is that although Bergson seeks the primordial phenomenon of time, he takes the wrong approach to bringing it to light. Bergson thinks that "time understood in the common way is really space" (BT 17/18) and that it is possible to uncover temporality by rigorously distinguishing time—that is, duration—from space. For Heidegger, however, it is not enough merely to distinguish the experience of time from the way time is commonly represented through spatial images and metaphors. Bergson's attempt to show how what we call "time" is really duration projected onto space and his opposition to all spatial thinking about time actually blocks access to primordial time, according to Heidegger. This is because it is not just the spatial representations we habitually use to think and speak about time that cover up temporality. Instead, as Heidegger argues in division two of *Being and Time*, originary temporality is concealed by the way Dasein "temporalizes" in the mode of "circumspect taking care" through which it interprets time. Heidegger thus mentions Bergson again at a pivotal point, in the transition to the chapters of division two in which he finally offers his account of how time, as it is ordinarily understood and traditionally conceived, is derived from temporality.[15] Although it differs in fundamental ways from originary temporality, Heidegger says, "the time 'in which' objectively present things come into being and pass away is a genuine phenomenon of time; it is not an externalization of a 'qualitative time' into space, as Bergson's interpretation of time—which is ontologically completely indeterminate and insufficient—would have it" (BT 318/333). This suggests that Bergson's neglect of the question of being is responsible for his view that the ordinary concept of time is illegitimate. Heidegger will argue, in the final chapter of *Being and Time*, that the concept of time is genuinely derived from ecstatic temporality, and therefore it is justified within certain limits.

Although Heidegger ultimately postpones any "critical discussion of Bergson's concept of time and other present-day interpretations of time" (BT 410n/433n), we can see that his dismissal of Bergson rests on three main complaints. First, according to Heidegger, Bergson tries but fails to clarify originary temporality with the concept of duration because he relies too heavily on the Aristotelian definition of time. By characterizing duration as "qualitative succession," Bergson merely turns Aristotle's concept on its head. This is evident from the fact that Bergson provides an analysis of number as a prelude to his discussion of duration, which supposedly has nothing to do with number. Bergson thinks that primordial time is a different kind of succession from the time that is measured, a succession of conscious states flowing into one another. As Heidegger will argue, however, primordial time is not a succession at all, but an ecstatic openness that grounds the being of Dasein. The idea of a "succession of nows," whether that succession is defined as qualitative or quantitative, belongs to the ordinary understanding of time. Contrary to Bergson, duration is not primordial time, but only one of the characteristics of "world time" (*Weltzeit*), which is the temporality of circumspect taking care. Bergson thus gets no closer to the root of time with his concept of duration than Aristotle does with his concept of time. To uncover the origin of time, according to Heidegger, it is necessary to rethink time on the basis of the existential structures of Dasein, as opposed to traditional, Aristotelian categories like quality and quantity, which are more appropriate to beings fundamentally different from us.

Heidegger's second complaint is that Bergson's attempt to radically rethink time suffers, like similar contemporary movements in philosophy, from a neglect of the question of being. As much as Bergson wants to break away from the philosophical tradition, he still employs traditional concepts—such as life, person, and consciousness—without considering their ontological presuppositions. In this respect, he belongs in the company of Dilthey, Husserl, and Scheler, who sense the need to think more radically about "human" existence but do not engage in the kind of ontological inquiry Heidegger views as necessary. Like these thinkers, Bergson challenges the reification of consciousness, but he does not call the being of consciousness into question. Even though he attempts to show how life, and especially conscious life, breaks the mold of thinghood, his thinking remains rooted in traditional ontology. His characterizations of consciousness or the person are mostly negative: it is not a mere thing, not extended, not divided into parts, not mechanical, and not predictable. For Heidegger, Bergson does not provide an adequate positive characterization of this kind of being, and, along with the other philosophers of life, he misses the phenomenon of being-in-the-world.

Finally, Heidegger argues that not only does Bergson fail to clarify originary temporality, but he also overstates the case against the ordinary concept of time. Contrary to Bergson's claim that common representations of time simply confuse it with space, according to Heidegger they refer to a genuine phenomenon, "now-time" (*Jetztzeit*). The ordinary understanding of time as a succession of nows in which everything occurs may not get to the root of temporality, but neither is it merely an illusion that we fall into as a result of spatial thinking. Instead, this concept of time is legitimate as a "natural" interpretation of temporality. The time in which natural processes and historical events occur is not originary temporality, Heidegger argues, but it is a phenomenon that arises from it. Thus, in his view, while Bergson does not go far enough in his search for primordial time, he goes too far in his rejection of the ordinary understanding of time. Moreover, thanks to his preoccupation with disentangling temporality from spatiality, Bergson misses the most crucial issue. What is ontologically decisive about the ordinary understanding of time, Heidegger thinks, is not space, but presence.

3. From Being-in-the-World to Temporality

So, what *is* time, according to Heidegger? As he had already emphasized in *The Concept of Time*, this question presupposes that time is a *being* or entity of some kind. But what it means *to be*, or the meaning of being in general, is precisely the question. Thus, time must be approached anew, in a way that challenges our ontological assumptions. In this lecture, he ultimately displaces the question "What is time?" by asking "Who is time?" and "Am I my time?" (CT 22/125). The following year, Heidegger concludes his *History of the Concept of Time* lecture course: "*Not 'time is' but 'Dasein qua time temporalizes its being*" (HCT 319/442). Following these initial attempts to reformulate the question of time, in *Being and Time*, Heidegger aims to think more radically about time by suspending the assumption that time is a being of some kind and asking about what it means to be temporal.

In his brief discussion of temporality in the introduction to *Being and Time*, Heidegger notes that time has traditionally been employed as a standard for distinguishing between two fundamentally different kinds of entity, " 'Temporal' beings (natural processes and historical events)" and " 'atemporal' beings (spatial and numerical relationships)" (BT 18/39/18). From the ancient Greeks until today, things that come into being and pass away have been understood as being *in time*, and things that are constant and unchanging as *timeless*, in other words "extra-" or "supra-temporal." In this way, time has been used to distinguish between two different regions of beings, those "within time" and those "outside" or "beyond" it.[16] Yet, Heidegger argues: "How time comes to have this distinctive ontological function, and even with what right precisely something like time serves as such a criterion, and most of all whether in this naïve ontological application of time its genuinely possible ontological relevance is expressed, has been neither asked nor investigated up to now" (BT 18/39/18). In fact, for Heidegger, such questions cannot be answered without a better understanding of the meaning of *temporality*. And temporality, interpreted as primordial time, must not be uncritically identified with being "in time" or defined on the basis of the traditional concept of time. Instead, temporality has to be approached through a preparatory analysis of Dasein, the kind of being that is not only in time, but "temporalizes," and this is the task of division one.

Heidegger's deepest concern with Bergson's thinking does not become clear until his discussion of the concept of time in the final chapter of division two. Only then, with his account of how time as a "succession of nows" is derived from the temporality of Dasein, does he clarify how Aristotle's investigation of time is definitive for the philosophical tradition up to and including Bergson. Our ordinary understanding of time is not the result of a confusion of duration with space, for Heidegger, but a natural interpretation of temporality in the mode of "making-present." Unfortunately, because Heidegger's account of temporality is built on a network of ontological structures and terminology developed in division one, any attempt to jump directly into his discussion of the relationship between originary temporality and the traditional concept of time is likely to be frustrated. Contrary to time, which is completely familiar to us on the basis of everyday experience, temporality is a phenomenon that, according to Heidegger, requires extensive preparation and great patience to understand. For the purpose of reserving our most careful attention for the final chapter of *Being and Time*, however, we can characterize Heidegger's path from the existential analysis of Dasein in division one to the temporal interpretation of Dasein in division two as a movement from the basic ontological constitution of Dasein, *being-in-the-world*, to the condition for the possibility of its unity, *temporality*.

Heidegger's analysis of being-in-the-world is at once a critical response to an interpretation of being that he attributes to the philosophical tradition as a whole and an attempt to provide a positive characterization of phenomena that are hidden by that interpretation. The orientation that Heidegger challenges will ultimately be revealed in Kant, Descartes, and Aristotle, but we have already seen how he identifies it in contemporary thinkers. According to Heidegger, what Bergson, Scheler, Husserl, and Dilthey try to do with their concepts of life, the person, and consciousness is to think about what fundamentally distinguishes human beings from other kinds of being. In his view, their failure to do so reveals a hidden truth about the limitations of the entire philosophical tradition going back to the ancient Greeks. The truth is that the being to which all these concepts refer has always been understood inappropriately in terms of "presence-at-hand" or "objective presence" (*Vorhandenheit*), a mode of being that characterizes whatever exists in nature. Heidegger attempts to demonstrate this understanding of being at work and point out its limitations through his analysis of Dasein as existing in another sense altogether. As opposed to the traditional understanding of existence as objective presence—to exist means to be present-at-hand (*vorhanden*)—Heidegger proposes that to *exist* means to be concerned about one's own being, thus to understand being in some way or another and to understand oneself in terms of possibilities. With respect to Dasein, "[t]he characteristics to be found in this being are thus not present 'attributes' of an objectively present being which has such and such an 'outward appearance,' but rather possible ways for it to be" (BT 41/42), and "[t]his being does not and never has the kind of being that is merely objectively present within the world" (BT 42/43). Accordingly, the aim of Heidegger's existential analytic of Dasein is to bring to light the characteristics and basic constitution of this being in such a way that it can no longer be confused with anything objectively present.

Although philosophers are quick to conclude that what distinguishes human beings from other kinds of being is that we are not just natural objects, but *knowing subjects*, Heidegger claims that a more fundamental difference, and one universally overlooked by philosophers, is the way that Dasein inhabits a *world* (*Welt*). For that reason, Heidegger uses the term "being-in-the-world" to designate the basic ontological constitution of Dasein. To be precise about what distinguishes Dasein from other beings, we first need to recognize the difference between the way something is objectively present in space, like a desk in a room, and the way we inhabit our environment (*Umwelt*). In Heidegger's view, a piece of furniture is "innerworldly" (*innerweltlich*), or objectively present in the world, but essentially "worldless" (*weltlos*), or indifferent to other beings. We, on the other hand, are "worldly" (*weltlich*) in the sense that our existence is defined by our worldly concerns, our dealings with the things we encounter in the world, and our disclosure of the context in which they have meaning to us. On the one hand, we are ontologically distinct from the objects surrounding us, which we understand in terms of our own possibilities, but on the other hand, we are nothing like an isolated, worldless subject of the sort that philosophers have traditionally identified with the acquisition of knowledge. For such a subject is commonly understood as something no less objectively present—as a peculiar sort of *thing* with its own distinctive properties—than the objects of nature. Yet, Heidegger says, "Dasein understands itself ontologically—and that means also its being-in-the-world—initially in terms of *those* beings and their being which it itself is *not*, but which it encounters 'within' its world" (BT 58–59/58). For Heidegger, being-in-the-world is a phenomenal unity prior to the very distinction between subject and object, which is at the root of the problem of knowledge. As a perceiver and a knower, he argues, "Dasein does not first go outside of the inner sphere in which it is initially encapsulated, but, rather, in its primary kind of being, it is always already 'outside' together with some being encountered in the world already discovered" (BT 62/62). Heidegger initially clarifies this in terms of the being of the objects that we encounter in our everyday dealings, the "readiness-to-hand" or "handiness" (*Zuhandenheit*) of useful things, but he ultimately explains it on the basis of temporality as an original "outside of itself" in the directions of the future, past, and present.

Without going very deeply into Heidegger's famous analysis of the "worldliness of the world," there are several points that will be important for understanding his discussion of "world time" (*Weltzeit*) in the final chapter of *Being and Time*. It is crucial that the "world" of being-in-the-world is, on Heidegger's analysis, not a totality of objectively present (*vorhanden*) beings—not the natural world of modern science—but a totality of *references*. Heidegger demonstrates this by distinguishing between the presence-at-hand of a thing with physical properties and the readiness-to-hand of a piece of equipment understood as something used in order to accomplish a task. He also describes how the "in-order-to" (*um-zu*) implies an intentional network including what that equipment is useful for, how it functions together with other equipment, and how its relevance is grounded in the possibilities "for-the-sake-of-which" (*worum-willen*) Dasein employs it. For any individual thing to show up as useful or useless, he argues, a referential totality—a world—must be presupposed in terms of which it receives its meaning and value. In his most famous example, the handiness of a hammer cannot be understood as the property of an isolated object apprehended by a knowing subject, but only with reference to nails, boards, the finished product, its purpose, and so on. Insofar as handy (*zuhanden*) beings are always referred to other beings, they are characterized by not just presence, but also *relevance*. Likewise, the world as a totality of references is characterized above all by *significance*.

Heidegger clarifies this with his critical destruction of the Cartesian conception of the world as *res extensa*. According to Heidegger, Descartes thinks about the world in terms of corporeal or extended substance thanks to "a basic ontological orientation toward being as constant objective presence" (BT 94/96), which causes him to "*pass over* the phenomenon of the world, as well as the being of innerworldly beings initially at hand" (BT 93/95). With respect to physical objects, Descartes privileges extension as that which persists or remains constant through changes to a thing's properties. In Heidegger's view, this reveals an assumption about the meaning of being going all the way back to the ancient Greeks: that being means substance or *constant presence*. Descartes's supposedly radical break with scholastic thought and his efforts to place knowledge on a new foundation are thus confined within the ontological tradition defined by the privilege of presence. Moreover, because even the most radical critics of modern philosophy and science do not challenge this privilege, they fail to distinguish the ontological character of those "innerworldly" things that are closest to us, as well as that of "worldly" beings such as ourselves, from objective presence. For Heidegger, Dasein cannot be understood in terms of a particular kind of substance, *res cogitans*, and the world cannot be understood in terms of spatial extension, which obscures its significance.

In Heidegger's analysis of being-in-the-world, he aims to avoid not only confusing the world with space, but also confusing Dasein with the "knowing subject" of modern philosophy. To that end, he attributes to Dasein a particular kind of being, "concern" or "taking care" (*Besorgen*), and a particular kind of seeing, "circumspection" (*Umsicht*), that indicate an everyday way of *dealing with* things, as opposed to knowing or perceiving them in a disinterested manner. "Taking care" encompasses a multitude of ways everyday Dasein can direct itself, for instance, "to produce, order, and take care of something, to use something, to give something up and let it get lost, to undertake, to accomplish," as well as the "*deficient* modes of omitting, neglecting, renouncing, [and] resting" (BT 57/56–57). "Circumspection," rather than "just looking" at the outward appearance of things, "guides our operations and gives them their specific certainty" (BT 69/69). As the kind of sight employed in dealing with useful things, it must be distinguished from the theoretical gaze. However, just as being-in-the-world is prior to the distinction between subject and object, for Heidegger, "circumspect taking care" (*umsichtig Besorgen*) is a mode of being prior to any distinction between theory and practice. It is active, engaged, and even absorbed in dealing with useful things, looking around or past them to put them to use most effectively. When Heidegger later attempts to explain how the concept of time is derived from Dasein's temporality, he begins with how time shows up in circumspect taking care before it is conceptually grasped or even measured.[17]

It is important to understand what Heidegger means by "care" (*Sorge*), because when he finally turns to temporality in division two of *Being and Time*, he introduces it as "the ontological meaning of care" (BT 309/323ff). Care is Heidegger's answer to other traditional definitions of the human being as a rational animal, a composite of mind and body, and so on. Appearing near the end of division one, it represents a "structural condensation" of three ontological characteristics of Dasein: "existentiality, facticity, and falling prey" (BT 185/191).[18] *Existentiality* means that Dasein "always has understood itself and will understand itself in terms of possibilities" (BT 141/145). In contrast to any objectively present thing, Dasein is always "ahead-of-itself," *projecting* itself on what it *can be*, its "potentiality-for-being" or "ability-to-be" (*Seinkönnen*). Likewise, Dasein is always "already-in-a-world," *thrown* or delivered over to the fact that it is and to whatever it *has to be*. *Facticity* means that Dasein's particular thatness shows up in its "state of mind" or "attunement" (*Befindlichkeit*), or that beings with the character of Dasein "find themselves in their thrownness, whether explicitly or not" (BT 132/135). Finally, Dasein is always "together-with-beings" and absorbed in those that are closest at hand. According to Heidegger, *falling prey* or *entanglement* (*Verfallen*) thus refers to being "absorbed in being-with-one-another" and "lost in the publicness of the they" (BT 169/175–6). As such, Dasein usually understands itself in terms of its everyday concerns and whatever "one does" and does not do. Heidegger unites the "existentials" of understanding, attunement, and falling prey under the rubric of care to indicate the inseparability of "being-ahead-of-oneself-already-in (the world) as being-together-with (innerworldly beings encountered)" (BT 186/192). With this, Heidegger distinguishes the "articulated structural totality" of Dasein's being from anything like a unidirectional stream of experiences. Care is a tripartite structure constituted by Dasein's "ahead of itself," its "already in the world," and its "together with beings."

By the end of division one of *Being and Time*, Heidegger seems to have gotten to the bottom of the being of Dasein. A "thrown projection" (*geworfene Entwurf*) that lets beings be discovered in their truth, Dasein is the "disclosedness" (*Erschlossenheit*) of the world, which, going back to the Greek *aletheia*, is "the most primordial phenomenon of truth" (BT 212/220). But what about this "primordial" (*ursprünglich*)? Heidegger asks: Does the ontological structure of care amount to a "*primordial* interpretation" of Dasein? Or is the structure of care grounded in a more primordial phenomenon? According to Heidegger, a truly primordial interpretation will have to shed light on "the being of Dasein in its possible *authenticity* and *wholeness*" (BT 223/233). However, there seems to be an impasse: because Dasein is always ahead of itself, it is a kind of being that is never given as a whole. Even at the moment of death, the being of Dasein is not complete, but rather it *ceases to be*. What is needed is an interpretation of Dasein that distinguishes its "being a whole" from that of anything objectively present, while also considering what constitutes authentic existence, as opposed to the average, everyday mode of being examined so far. A primordial interpretation must show how to understand Dasein as a whole—not in a state of completion, but as a "potentiality-for-being-whole"—*in terms of time*. As Heidegger says:

> The primordial ontological ground of the existentiality of Dasein . . . is *temporality*. The articulated structural totality of the being of Dasein as care first becomes existentially intelligible in terms of temporality. . . . We must go back and free the ontological structures of Dasein already gained with regard to their temporal meaning. (BT 224/234)

In other words, a primordial interpretation of Dasein will be a *temporal interpretation* of the tripartite care structure and its constituent parts, each of which is grounded in an "ecstasy" of temporality.

4. From Temporality to Time

Temporality is thus the key to Heidegger's primordial interpretation of Dasein, the kind of being traditionally designated as the subject, ego, mind, spirit, or soul—and by Bergson as consciousness or psychic life. However, temporality is not a "lived experience," for Heidegger, but the ontological condition for the possibility of the unity of care, the counter-possibilities of authentic and inauthentic existence, and our ordinary understanding of time. What, then, is temporality? This question faces the same problem that led Heidegger to avoid the question "What is time?" To say "temporality is . . ." presupposes that temporality is a being of some kind.[19] In §65 of *Being and Time*, finally making a decisive transition to the temporal interpretation of Dasein, Heidegger writes:

> Temporality "is" not a being at all. It is not, but rather temporalizes itself. Nevertheless, we still cannot avoid saying that "temporality 'is' the meaning of care." . . . Temporality temporalizes, and it temporalizes possible ways of itself. These make possible the multiplicity of the modes of being of Dasein, in particular the fundamental possibility of authentic and inauthentic existence. (BT 314/328)

When we think about temporality, as opposed to time, we need to understand it as a "how" rather than a "what." More specifically, temporality is how Dasein can exist "ahead of itself" at the same time that it is "already in" a factical situation "together with" other beings. That is to say, "[t]emporality makes possible the unity of existence, facticity, and falling prey and thus constitutes primordially the wholeness of the structure of care" (BT 313/328). Originary temporality is thus the *condition for the possibility of care*, its transcendental foundation or ground (*Grund*). This is why Heidegger calls temporality the "meaning" (*Sinn*) of care: it is that without which care would be neither possible nor intelligible. As such, the structure of originary temporality will reflect and illuminate the tripartite structure of care, and it will be reflected, albeit dimly, in the way time is ordinarily understood.

Before detailing how temporality makes care—and thus the being of Dasein as such—possible, Heidegger shows how authentic existence can be understood in terms of temporality as *anticipation*. Beginning with the question of how to understand Dasein in its authenticity and wholeness, Heidegger presents his famous and controversial analysis of death. Dasein is a being that is *finite* and understands itself as such. What is important, for the purpose of Heidegger's temporal interpretation of Dasein, is not just that it comes to an end, or dies, but that it directs itself toward its end—that Dasein is "being-towards-death." For the most part, Heidegger argues, Dasein's stance toward its own death is one of evasion: we "flee" from death by regarding it as something that happens to others, or something that will happen in the distant future. In this way, "death is understood as an indeterminate something which first has to show up from somewhere, but right now is *not yet present* for oneself," and "[d]ying, which is essentially and irreplaceably mine, is distorted into a publicly occurring event which the they encounters" (BT 243/253). However, in contrast to this *inauthentic* attitude toward death, Dasein is capable of an *authentic* being-towards-death. In contrast to the view of death as an actual event at some indeterminate point in one's future, death "must be understood *as possibility*, cultivated *as possibility*, and *endured as possibility* in our relation to it" (BT 250/261). Moreover, we must direct ourselves toward the possibility of our own death not through "expectation" (*Erwarten*), merely awaiting its actualization, but through "running ahead" or "anticipation" (*Vorlaufen*).[20]

According to Heidegger, "[a]nticipation shows itself as the possibility of understanding one's *ownmost* and extreme potentiality-of-being, that is, as the possibility of *authentic existence*" (BT 252/263). As opposed to evading death, anticipation means owning up to it, understanding ourselves and our possibilities as finite, limited by the certainty and indefiniteness of life's coming to end. Anticipation "*frees*" Dasein "*for* death," according to Heidegger, and "liberates one from one's lostness in chance possibilities urging themselves upon us, so that the factical possibilities lying before the insuperable possibility [of death] can first be authentically understood and chosen" (BT 253/264). Anticipation, as he understands it, discloses us to ourselves, enables us to choose who we will become, helps us pay heed to the voice of conscience, leads us to accept responsibility for our choices, and points us toward authentically "being a self." To provide a full view of authentic existence as he sees it, Heidegger couples anticipation with *resoluteness* (*Entschlossenheit*), "a self-projection upon one's ownmost being-guilty in which one is reticent and ready for anxiety" (BT 288/301). Resoluteness discloses not only Dasein's "ownmost possibilities," but also the situation that affords them, "[f]ar removed from any objectively present mixture of the circumstances and accidents encountered" (BT 287/300). As opposed to the variety of experiences in which we are irresolute, absorbed in worldly concerns and lost in the "they," Heidegger claims, "*[t]emporality is experienced as a primordial phenomenon in the authentic being-whole of Dasein, in the phenomenon of anticipatory resoluteness* (BT 291/394). Thus, authentic temporality *is* an experience of a certain kind, and one that allows us to catch sight of a deeper ontological structure, but it has nothing to do with the "flow of time" as commonly conceived.

Finally, with Heidegger's account of temporality as the meaning of care (§65), the difference between originary temporality and Bergsonian duration begins to come to light. The origin of time is to be found not in ordinary time-consciousness, for Heidegger, but in the structure of temporality implied by *authentic* existence. Heidegger first describes the structure of originary temporality in a discussion that culminates in four theses:

> Time is primordial as the temporalizing of temporality, and makes possible the constitution of the structure of care. Temporality is essentially ecstatic. Temporality temporalizes itself primarily out of the future. Primordial time is finite. (BT 316/331)

The first thesis asserts that temporality "is" the meaning of care, but also that, strictly speaking, temporality is not a being at all, but a process or movement of "*temporalizing*" or "*temporalization*" (*Zeitigung*) that makes the being of Dasein possible.[21] Originary temporality can be understood as a complex movement that, as the condition for the possibility of care, is exhibited most clearly in authentic existence, that is, anticipatory resoluteness. First, a condition of anticipation is that Dasein be able to "come toward itself . . . in its ownmost possibility," and Heidegger interprets this "*coming toward*" (*zukommen*) as "the primordial phenomenon of the *future* (*Zukunft*)" (BT 310–11/325).[22] In anticipation, the future is not something "not yet actual" that comes toward us, but rather "the coming in which Dasein comes *toward itself*" (BT 311/325, emphasis added). Second, resoluteness requires Dasein to "take over thrownness," to be "in the *way that it always already was*," or to be as "having-been" (*gewesen*) (BT 311/325–6). In this "*having-been*" (*Gewesenheit*), which is the primordial phenomenon of the past, Dasein's *coming-toward* what it will be is doubled by a *coming-back* (*zurück-kommen*) to what it already was. In this sense, having-been "arises from the future" as Dasein takes over its thrownness by projecting itself on its authentic possibilities.[23] Unlike the past as we typically think of it, "having-been" is not what has "passed by" and "ceased to be."[24] Third, anticipatory resoluteness involves acting, and "letting *what presences* [*Anwesenden*] in the surrounding world be encountered in action, is possible only in a *making* that being *present* [*Gegenwärtigen*]" (BT 311/326). The primordial phenomenon of the present is therefore the back-and-forth between the future and having-been that enables Dasein to "take hold of" or grasp things as belonging to its situation, the movement of *presencing* as such.

Heidegger thus defines temporality as the "unified phenomenon of the future that makes present in the process of having-been (*gewesend-gegenwärtigende Zukunft*)" (BT 311/326). We can now see how this phenomenon is implicit in the care structure: the future as coming toward indicates how Dasein is *ahead of itself*; having-been indicates how Dasein is *already in the world*; and the present indicates how Dasein is *together with beings*. As such, "[t]emporality makes possible the unity of existence, facticity, and falling prey, and thus constitutes primordially the wholeness of the structure of care" (BT 313/328). Heidegger argues that each of these ontological characteristics depends on the whole structure of temporality, but he also assigns a primary temporal meaning to each: existentiality is grounded in the future, facticity is grounded in having-been, and falling prey is grounded in making-present. He warns us, though, that when we speak of temporality, we must be careful not to confuse it with "all the meanings of 'future,' 'past,' and 'present' initially urging themselves upon us from the vulgar concept of time" (BT 312/326). These parts or tenses of time are reified versions of the movements of coming-toward, having-been, and making-present.

This brings us to Heidegger's second thesis, that temporality is "essentially ecstatic." He introduces the term "ecstatic" in an attempt to depart radically from the traditional interpretation of being as substance. In contrast to the Latin "*substantia*," meaning that which "stands under" or supports something, the Greek *ekstatikon* refers to something that "stands out" or is unstable. With all his talk about temporality as a "ground" for authentic existence, his aim in first interpreting Dasein in terms of *existence* and then reinterpreting its existential structures in terms of temporality is to distinguish its being from the objective presence of a substance or a subject. To this end, he calls temporality "the ἐκστατικόν [*ekstatikon*] pure and simple," explaining:

> *Temporality is the primordial, "outside-of-itself" in and for itself.* We therefore call the phenomena of future, having-been, and present the *ecstasies* of temporality. . . . its essence is temporalizing in the unity of the *ecstasies*. (BT 314/329)

The unity that Heidegger describes here is something entirely different from that of an external object, a conscious subject, or even a sequence of experiences in time. The ecstasies of temporality do not follow one another, but rather they comprise a unity that is ontologically prior to the ordinary understanding of time as a "succession of nows." Thus, the future and having-been are not later and earlier than the present—they cannot be understood as not yet and no longer present—but are coeval or "equiprimordial" (*Gleichursprünglich*) with it.[25] The characterization of temporality as ecstatic thus indicates a fundamental difference between primordial time and time as we ordinarily understand it, according to Heidegger: there is no primacy of the present.

In fact, although the ecstasies of temporality are equally originary, Heidegger's third thesis is that temporalization occurs "primarily out of the future." Like the first thesis, this follows from the preparatory account of authentic temporality as anticipatory resoluteness. Although authenticity is only possible on the basis of the unity of the ecstasies, insofar as it is distinguished by *anticipation*, the future is primary for Heidegger. In his view, "[p]rimordial and authentic temporality temporalizes itself out of the authentic future, and indeed in such a way that, futurally having-been, it at first arouses the present" (BT 314/329). One indication of this is the way that authentic Dasein "comes toward itself" so that it always "has time" for what matters, and in this sense is fully present, whereas inauthentic Dasein "has no time." Thus, while the past and the future have traditionally been characterized negatively in relation to the present—as no longer and not yet present—Heidegger challenges this by suggesting that having-been and making-present must be understood positively in relation to the authentic future. Not only is there no primacy of the present in the primordial phenomenon of time, but rather the most primordial phenomenon of time, anticipatory resoluteness, displays a certain primacy of the future insofar as Dasein understands itself in terms of possibility.[26]

Heidegger's final thesis concerning originary temporality may be his most surprising: it is finite. Despite how it sounds, this does not mean that time comes to an end, that it terminates or stops at some point, but rather that the temporality of Dasein is defined by being-toward-death. Thus, what is finite is the *temporalizing* of temporality, the movement by which Dasein is what it can be and has been in such a way as to take hold of the situation. Death, as the ownmost possibility of Dasein, provides a closure to temporality that corresponds to its ecstatic openness. The future as we ordinarily understand it is full of endless possibilities, and time marches on after our demise, but authentic temporality means coming toward ourselves from the one possibility, our own death, that brings all the others to a close. Dasein has to understand itself as finite, as coming to an end, to temporalize authentically. However, if primordial time is finite in this way, then why do we think of time as endless or infinite? According to Heidegger, our tendency to overlook the finitude of authentic temporality is due to "the constant intrusion of the vulgar understanding of time" (BT 315/330). This understanding, which has been presupposed in all philosophical debates concerning time, he concludes, must arise from inauthentic temporality.

Heidegger defends his theses about originary temporality as "the future that makes present in the process of having-been" on the basis of his account of authentic existence, in which it shows up most clearly. However, originary temporality is what makes care in general, and therefore also inauthentic existence, possible. It is crucial, in Heidegger's view, that the primordial phenomenon of time be distinguished from the concept of time as we ordinarily understand it, including the "past," "present," and "future." As opposed to the interpretation of temporality he has developed, the ordinary concept arises from an inauthentic understanding of temporality: "Since Dasein understands itself initially and for the most part inauthentically, we may suppose that the 'time' of the vulgar understanding of time indeed presents a genuine phenomenon, but a derivative one" (BT 312/326). As he will argue in the final chapter of *Being and Time*, this ordinary understanding of time is not an illusion or an illegitimate concept because it arises from temporality. Nevertheless, this concept of time reflects originary temporality in a distorted way. The temporality that is ecstatic, finite, and futural—the primordial phenomenon of time—is nothing present, but is an "outside of itself" or "openness" that makes presence possible for Dasein. With this interpretation of temporality in hand, fleshed out by temporal interpretations of everydayness and historicity, Heidegger will attempt to show how exactly the concept of time is derived.

According to Heidegger, our ordinary understanding of time arises not from the authentic (*eigentlich*) temporality of anticipatory resoluteness, but from the inauthentic (*uneigentlich*) temporality of average, everyday life. He elaborates on this distinction in his temporal interpretation of Dasein's understanding (§68a), where he describes authentic and inauthentic versions of each of the three ecstasies of temporality. First, as essentially ahead-of-itself, Dasein "comes toward itself" existentially. In so doing, it either "lets itself come toward itself as its ownmost potentiality-of-being" (BT 322/336–7), anticipating death, or "projects itself upon what can be taken care of" and "does not come toward itself primarily in its ownmost, nonrelational potentiality-of-being [i.e., death], but *awaits this* heedfully" (BT 322/337). As a way of temporalizing, and a common one at that, the inauthentic future is thus defined by awaiting (*Gewärtigen*). Second, as essentially together-with innerworldly beings, Dasein brings itself into a situation, either taking hold of it in action or losing itself in worldly concerns. The authentic present, which Heidegger calls the "blink of an eye" or "moment" (*Augenblick*), "is not only brought back from dispersion in what is taken care of closest at hand, but is held in the future and having-been" (BT 323/338). The moment, he says, "must be understood in the active sense as an ecstasy," as "rapture *held* in resoluteness," and it "can *in principle not* be clarified in terms of the *now*" (BT 323/338). The moment is not a characteristic of the time "in which" things happen, but "lets us *encounter for the first time* what can be 'in time' as something at hand or objectively present" (BT 323/338). Although he has already used the term making-present (*Gegenwärtigen*) for the ecstasy of the present, he now gives it a more specific sense of the inauthentic present, the present of irresoluteness and absorption in taking care, in contrast with the moment. Finally, Dasein, as essentially already-in the world, is always coming back to what it has been, to its facticity. Heidegger calls authentic having-been "retrieval" or "repetition" (*Wiederholung*), "a coming back to the ownmost self thrown into its individuation," that which "makes it possible for Dasein to be able to take over resolutely the being that it already is" (BT 324/339). In contrast, inauthentic having-been is forgetting (*Vergessenheit*), which is not "just a failure to remember," but rather "a backing away from one's ownmost having-been in a way that is closed off from oneself" (BT 324/339). Inauthentic temporality can thus be described, as opposed to the authentic temporality of anticipatory resoluteness, as "awaiting that forgets and makes present." What makes this mode of temporality inauthentic is that it "closes off one's authentic potentiality-of-being, and is thus the existential condition of the possibility of irresoluteness" (BT 324/339). As we will see in Heidegger's discussion of world time, this inauthentic mode of temporality is also, in his view, the condition for the possibility of our ordinary understanding of time.

Having distinguished between authentic and inauthentic temporality, Heidegger returns to the phenomenon of everyday being-in-the-world to show how temporality makes it possible. More specifically, he wants to show how ecstatic temporality is the condition for the possibility not just of care (*Sorge*), but also of everyday taking care (*Besorgen*). According to his analysis in division one, the world is first and foremost a totality of relevance, wherein Dasein encounters beings in the mode of circumspect taking care, and they show up as handy (or not). Dasein is usually absorbed in taking care, and Heidegger claims that such absorption is grounded in a particular mode of temporality, the "making present that awaits and retains" (BT 337/384). For something to be handy, it must appear in the light of a project of Dasein, its relevance being determined by the possible results that Dasein "awaits" and the means of producing them that Dasein "retains." Here, awaiting (*Gewärtigen*) is similar to expectation (*Erwarten*) in the sense that we employ a tool, for example, in the expectation that it will perform its function, but Heidegger adds the caveat that awaiting occurs even in the absence of an explicit, consciously apprehended goal. Retaining (*Behalten*) refers less to an operation of memory than the capacity for encountering useful things in their context, as connected with other things to which they refer rather than as isolated objects; in fact, it even involves a particular kind of forgetfulness: the self-forgetfulness that is necessary "to be able to 'really' get to work, 'lost' in the world of tools and to handle them" (BT 337/354). Thus, for Heidegger, "being relevant constitutes itself in the unity of awaiting and retaining in such a way that the making present arising from this makes the characteristic absorption in taking care in the world of useful things possible" (BT 337/354). Such absorption is only possible if the things encountered are familiar and inconspicuous, their relevance being taken for granted. In division one, Heidegger described how the handiness of things is invisible to everyday Dasein until something becomes conspicuously un-handy, interrupting the flow of work. Now he explains why the temporality of taking care must be understood as an ecstatic unity: "If heedful dealings were simply a succession of 'experiences' occurring 'in time' . . . letting a conspicuous, unusable tool be encountered would be ontologically impossible" (BT 338/355). With traditional ontology, we face the dilemma of treating conspicuousness as a special property of certain objects or a merely subjective appearance—in any case, as something present. The need to understand these phenomena on the basis of ecstatic temporality becomes clearer in the case of something that goes missing: we can only discover that something is missing if we expect it to be somewhere that it is not, and such an expectation is, in Heidegger's terms, the making-present of an absence or lack of something whose appearance is awaited and, insofar as we know what to look for, retained. Circumspect taking care thus needs to be interpreted temporally as grounded in the making-present that awaits and retains.

5. The Ordinary Understanding of Time

Heidegger's attempt to radically rethink time has led him to describe temporality as the meaning of care, that is, the condition for the possibility of the kind of being that is "ahead of itself," "already in the world," and "together with innerworldly things." Interpreting it in this way, he cautions against claiming that temporality *is* anything whatsoever, which would imply that it is an entity of some sort. Instead, we should say that temporality *temporalizes* or that Dasein temporalizes in order to indicate that temporality is an ontological structure that makes various modes of being possible. If temporality is primordial time, then where does this leave us with respect to understanding time? In the final chapter of *Being and Time*, Heidegger notes that we commonly speak of historical events as taking place "within time." What exactly is this time in which not only historical but also natural occurrences are thought to occur? What is the meaning of "within-timeness" or "intratemporality" (*Innerzeitigkeit*), and what does it have to do with the temporality of Dasein? These questions bring Heidegger closer to clarifying the relationship between originary temporality and time as we ordinarily understand it.

With reference to the structure of originary temporality, Heidegger has gone more deeply into the ontological characteristics of everydayness and deepened his analysis of Dasein with an account of historicity. To complete the temporal interpretation of Dasein, Heidegger intends to show "how and why the development of the vulgar concept of time comes about in terms of the temporally grounded constitution of being of Dasein taking care of time" (BT 386/405). Building on his analysis of being-in-the-world, "taking care of time" (*Besorgen von Zeit*) designates the many ways that, "before all thematic investigation, Dasein 'reckons with time' and orients itself *according to it*" (BT 385/404). Although temporality is not an entity that we encounter in the world, we speak of ourselves as "having" or "not having" time, "taking" time, "losing" time, and so forth. For Heidegger, this shows that Dasein *expresses* its temporality and *interprets itself* in temporal terms prior to having a conceptual understanding of time. Thus, to demonstrate how the concept of time is related to originary temporality, he examines the "time taken care of" (*besorgte Zeit*), which is the way time first shows up for Dasein as a being-in-the-world.

In the final chapter of *Being and Time*, Heidegger continues to engage with Bergson, following him in some respects and diverging in others. With Bergson, Heidegger argues that what we usually think of as time is a concept derived from a more primordial phenomenon. Our ordinary understanding of time "arises" or "springs from" (*springt aus*) originary (*ursprünglich*) temporality, according to Heidegger, and it does so in such a way that the structure of temporality is hidden. Time does not have the same characteristics as the temporality of Dasein, but it does reflect the structure of temporality in somewhat distorted form. The concept of time, as Bergson already recognized, both conceals and reveals the primordial phenomenon from which it is derived. However, against Bergson, Heidegger denies that the concept of time is a spatial representation of duration. Bergson's strategy in *Time and Free Will* was to distinguish our concrete, lived experience of time from abstract representations of it. In his view, duration is "immediately given," and all we need in order to see it more clearly is to stop thinking about time in spatial terms. To dispel the confusion of duration with extension, it is necessary to break the habit of representing time as a homogeneous medium analogous to space. This is an illegitimate concept, in Bergson's view, and one that is responsible for many of the apparently insoluble problems of metaphysics. On the contrary, Heidegger aims to establish that the time in which things come into being and pass away is not a spatialization or "externalization" of time, but a "genuine phenomenon" (BT 318/333). In opposition to Bergson's attempt to rethink time, Heidegger argues that the crucial issue, and the source of the philosophical difficulties surrounding time, is not space, but presence.

Heidegger's discussion of the time taken care of is framed by two metaphysical debates that he has avoided so far. The first is the question of whether time is objective or subjective, an issue over which there has been "a remarkable vacillation" (BT 386/405) in the history of philosophy. The question arises for Aristotle, who wonders whether time is implicit in motion itself or depends on the ability to perceive motion and count, leading him to ask about the connection between time and the soul. If time is a measure of motion, "something counted," then how could it exist independently of the mental activity of counting?[27] The second question, which is closely connected with the first, is whether time exists—"whether and how time has any 'being' " (BT 386/406)—or is merely an illusion. This question goes back to the pre-Socratics, especially Parmenides and Zeno of Elea, but it is posed definitively by Aristotle, who asks how time can exist if it is composed of nonexistent parts: the past, which *is no longer*, and the future, which *is not yet*. In much the same way that Bergson attempts to resolve the problem of freedom by revealing the misconceptions from which it arises, Heidegger's aim is not to answer these questions directly, but to focus on the ontological presuppositions that make them perplexing. In the final chapter of *Being and Time*, Heidegger tries to show that philosophers have failed to come to grips with *the being of time* because they have assumed that whatever exists must be objectively present.

Heidegger's demonstration of how the ordinary concept of time arises from originary temporality takes several steps, beginning with four characteristics of time so familiar that they escape most philosophers' attention and gradually showing how these characteristics are obscured as time comes to be not only measured, but also grasped conceptually as something capable of being measured. Because Heidegger is still in the process of working out his account of the derivation of time from temporality, the distinctions he makes at various stages are not always clear or consistent. Ultimately, though, he uses the expression "world time" (*Weltzeit*) for the way temporality shows up in circumspect taking care, and the "time of nature" (*Naturzeit*) or "now-time" (*Jetztzeit*) for the way world time is conceptually understood. Heidegger's aim at this stage is to show how the characteristics of world time are derived from Dasein's temporality, so that he can go on to show how the concept of time "levels down" world time. Against Bergson, this means that although our ordinary understanding of time is not primordial, it is still a legitimate way to conceive the time in which historical events and natural processes occur. Bergson is wrong, in Heidegger's view, to think of duration as primordial temporality and measurement as a projection of it into space. In his discussion of world time, Heidegger attempts to account for duration without giving it central importance and to offer an alternative account of what happens to temporality with the measurement of time.

On Heidegger's account, the four essential characteristics of world time are datability, spannedness, publicness, and significance, and they can all be understood with respect to temporality. Heidegger begins by emphasizing that Dasein both expresses and interprets itself, even prior to any conceptual understanding of time, in temporal terms. The first characteristic of world time, *datability*, involves the way that temporality first shows up in everyday discourse. Heidegger points to three basic expressions of time whose meaning is self-evident: "As taking care in calculating, planning, preparing ahead, and preventing, [Dasein] always already says, whether audibly or not: '*then* . . . that will happen . . . , '*now*' . . . that will be made up for, that '*on that former occasion*' failed or eluded us" (BT 387/406). The terms "now" (*jetzt*), "then" (*dann*), and "on that former occasion" or "formerly" (*damals*) are commonsense expressions of the temporality of Dasein, specifically "the mode of making present that awaits and retains" (BT 387/406). On Heidegger's account, the "now" is an expression of making-present, the "then" of awaiting, and the "on that former occasion" of retaining. Thus, each term corresponds to one of the ecstasies of temporality: the "now" expresses a certain taking hold, the "then" a coming toward, and the "formerly" a coming back. Moreover, the temporal horizons of "today," "later," and "earlier" are implicit in these expressions.

By means of this "seemingly self-evident relational structure" (BT 388/407), Dasein gives its projects and significant events a more or less definite "date-stamp." Prior to the specific dating of events with a calendar, Dasein takes care of things by saying, for instance, "now that" the soil is dry, "then when" the planting is done, and "formerly" when the crop failed. While the "now" is commonly conceived as a "point of time," in circumspect taking care, Dasein always refers beyond itself to occurrences in the world. Like the spatial expressions "here" and "there," the terms of datability are neither objectively present nor handy things, yet they are always already available to Dasein. This is so, according to Heidegger, because the "now" and its counterparts express the ecstatic openness of Dasein's temporality.

Heidegger's discussion of the relational structure of datability forges the first links between the way time is ordinarily understood and originary temporality. This structure shows how temporality is the condition for the possibility of our understanding of time, yet it is concealed by that understanding. "The making present that awaits and retains interprets *itself*," according to Heidegger, "because, in itself ecstatically open, it is always already disclosed to itself and can be articulated" (BT 388/408). However, what shows up in circumspect taking care is not the complete, articulated unity of ecstatic temporality, but a particular mode of it: "The making present that interprets itself, that is, what has been interpreted and addressed in the 'now,' is what we call 'time'" (BT 389/408). Although time is familiar and "immediately" grasped by us, as the expressions of datability attest, it does not follow that their source is immediately given. While originary temporality is the unity of the future, having been, and the present, what shows up in and for circumspect taking care are the familiar "then," "formerly," and "now," which "*stem from temporality and are themselves time*" and together amount to "the most primordial way of *giving the time*" (BT 389/408). This relational structure shows up even before the actual use of calendars or clocks, much less any conceptual understanding of time; therefore, it is more fundamental than the measurement of time.

The second essential characteristic of world time, which Heidegger calls *spannedness*, is closely connected to datability. The description of this aspect of time is Heidegger's only discussion of duration in *Being and Time*, yet curiously it includes no mention of Bergson. Heidegger begins not with a full-fledged concept of duration, but with the temporality of Dasein and how it is interpreted and expressed in taking care. Along with the "now" that expresses making present and the "then" that expresses awaiting, there is the interval between them. Implicit in the "then" is "not yet now," so the making present that awaits also expresses itself as "until then." The interval or "in between" (*Inzwischen*) shows up in our talk about *having time*, for instance, having time to harvest *until* the first freeze. As Dasein projects itself toward other "thens" in such an interval—"then, when the tractor is repaired," "then, when crop prices peak"—it expresses itself in terms such as "while" or "during" (*während*)—"while it is broken," "during the glut on the market." This structure also displays ecstatic temporality:

> The "duration" (*Dauern*) is articulated in the understanding of the "during" (*wahrend*) that awaits and makes present. This duration is again the time revealed in the *self*-interpretation of temporality, a time that is thus actually, but unthematically understood in taking care as a "span." The making present that awaits and retains interprets a "during" with a "span," only because in so doing it is disclosed to *itself* as being ecstatically *stretched along* in historical temporality, even if it does not know itself as this." (390/409, translation modified)

Dasein *stretches* from now to then, and a stretch or span of time is how it articulates the "during" of this project or "in between" these events. Stretching is characteristic of not only the interval, but also the "now" and "then" themselves. This is evident from the variability with which "now" is used—"now, in the evening" as opposed to "now, in the summer" or even "now, in the age of mass production." For circumspect taking care, every event is datable in relation to others, and by virtue of Dasein's temporal stretching out, time shows up in the interval between dated events and the duration of those events themselves. On this account, duration is not the structure of originary temporality, but a characteristic of the intervals opened up by it.

If datability indicates a primordial way in which time "is given," spannedness indicates how Dasein "allows itself" time prior to grasping it conceptually. Just as dating refers to events and beings in the surrounding world, stretching deals with Dasein's daily concerns, or with "what one does 'all day long'" (BT 390/409). The way Dasein passes a typical day, not being able to account for every interval, leaves time with gaps in it. Nevertheless, thanks to the ecstatic unity of temporality, the day does not fall apart, but hangs together as a span of time. This togetherness of world time is prior to the continuity of time as it is traditionally conceived, for in circumspect taking care, "Dasein never understands itself as running along in a continuously enduring succession of sheer 'nows' (BT 390/462/409). The spannedness of time may anticipate the way that time is conceived in terms of continuity, but it is more primordial than "the theoretical 'representation' of a continuous stream of nows" (BT 390/410). As Heidegger will argue when he turns to the concept of time, the continuity of time is based on the unity of temporality. Insofar as Bergson emphasizes continuity, this implies that his thinking remains at the level of the traditional concept.

With the difference between authentic and inauthentic temporality, Heidegger explains how, even prior to the measurement of time, we interpret it as something objectively present, a possession. Insofar as Dasein "allows itself" time, it seems as if we should always "have time," as long as Dasein is ecstatically open. But if time is not a being, then what sense does it make to speak of "having" or "not having" time? The problem here is similar to the one that Heidegger discusses in division one with regard to the idea of human beings "having an environment," a problem that, in his view, can only be solved by "defining the being of Dasein beforehand in an ontologically adequate way" (BT 58/58). In the case of time, he argues, "having no time" is characteristic of inauthentic existence, in which "[t]he irresolute person understands himself in terms of the events and accidents nearest by that are encountered in such a making present [that forgets] and urge themselves upon him in changing ways. Busily losing *himself* in what is taken care of, the irresolute person *loses his time* in them, too" (BT 391/410). In contrast, resoluteness implies stretching along "in the sense of the authentic, historical *constancy* of the self," such that the resolute person "never loses time and 'always has time,'" or more precisely, "'constantly' has time *for* that which the situation requires of it" (BT 391/410). "Having" time thus depends on how Dasein temporalizes, either losing itself in the "they," or coming back to itself in a way that discloses the situation in the moment. While Dasein "allows itself" time by virtue of its ecstatic openness, it is always possible to "lose" time in distraction or to "take" time to project itself authentically. Both modes of existence arise from ecstatic temporality and are prior to the concept of time as something that we possess but that constantly slips away.

Heidegger reveals an interesting affinity to Bergson when he highlights the way that saying "now" is more or less explicitly a social act. The first and second characteristics of world time, datability and spannedness, correspond to the existential structures of being-together-with innerworldly things and being-a-self. For Dasein "dates" according to what it takes care of and "stretches along" (both daily and historically) in a way that constitutes its identity. The third essential characteristic of world time, *publicness*, corresponds to being-with-others. Insofar as Dasein interprets and expresses its temporality in a "public, average" way as something shared with others, something "there" for everyone, time is "always already *made public*" (BT 391/411). This characteristic also arises from originary temporality, insofar as it is ecstatic and "opens up" Dasein in such a way that it is possible to say "now" and "then" with others and to understand what others mean by these expressions as well.[28] The publicness of time is evident in the way time is simultaneously understood both as one's own possession and as "public property." Indeed, the ability to orient ourselves according to time demands that "it must somehow be available for everyone" (BT 392/411). Heidegger reverses Bergson's order of explanation, however, when he argues that making public is a condition for the possibility of time reckoning and the measurement of time. It is not the case, in his view, that we represent time as a quantity, and, as a result, it is something public; on the contrary, time has to be interpreted as public to be measured in the first place.

Heidegger's analysis of time reckoning (*Zeitrechnen*) shows how circumspect taking care makes it possible for Dasein to measure time. The term "reckoning" traditionally refers to the calculation of the months, seasons, and years for purposes such as establishing religious holidays and predicting astronomical events. However, Heidegger claims that "the essence of taking care of time does *not* lie in the application of numerical procedures in dating" or, in somewhat more Bergsonian terms, "the quantification of time" (BT 392/412). More important in his view is how time reckoning exhibits the structure of care, especially facticity and thrownness. Indeed, the very notion of "everydayness" can be traced back to the way Dasein uses the sun to reckon with time. Thrown into a world where daylight alternates with darkness, Dasein can date events in relation to the rising and setting of the sun, which makes *sight* and thus circumspect taking care possible.

The light of the sun, its regularity, and its public availability make it handy for coordinating with others. Sunrise, noon, and sunset provide constant references for the "then, when . . ." of world time, serving as "distinctive 'places' " or positions to be used in dating, signs that appear daily with the sun's "regular recurring passage" (BT 393/413). By dating events according to positions of the sun, Dasein has a clock in the surrounding world, a clock that "is always already discovered with the factical thrownness of Dasein grounded in temporality" (BT 394/413). The day is a natural measure of time, but only because Dasein exists in the mode of making present that awaits and retains. This "natural clock" grounded in Dasein's temporality makes it possible to produce and use artificial clocks.

Using a clock to measure time thus arises from ecstatic temporality, but before Heidegger goes into detail about this, he outlines the fourth and final essential characteristic of world time, *significance*. The significance of time is evident in the way the "then, when . . ." of datability is always a "time for" (*Zeit zu*) something or other. The expression "[then,] when the sun rises, it is time for work," for instance, exhibits an understanding of time as appropriate for some activities—getting out of bed—and inappropriate for others. "Time for . . ." implies the "what for" or "in-order-to" (*um zu*) and the ultimate "for-the-sake-of-which" of worldly involvement, which Heidegger called "significance" in his analysis of worldhood in division one. It is because of this worldly character of significance that Heidegger calls the time taken care of "world time," a time that is "datable, spanned, and public and, having this structure, it belongs to the world itself" (BT 395/414). World time, which shows up in the way that temporality interprets and expresses itself in circumspect taking care, is the bridge between ecstatic temporality and the ordinary concept of time. For the way Dasein takes care of time in the form of reckoning makes it possible to measure time and ultimately to understand it as a succession of "nows."

Heidegger's discussion of world time leads to a phenomenological interpretation of time measurement (*Zeitmessung*) in relation to the temporality of Dasein. In opposition to Bergson's attempt in *Time and Free Will* to demonstrate that the time measured by clocks is a representation of duration that confuses it with space, Heidegger aims to provide a more ontologically adequate account of what makes such measuring possible. While maintaining the difference between originary temporality and ordinary time, Heidegger rejects the notion that to measure time is to "externalize" or project it into space (BT 318/333). On the contrary, for Heidegger, measuring time is an expression of Dasein's temporality: it is a way of temporalizing or, more specifically, making-present (*Gegenwärtigung*). Heidegger's analysis of time measurement demonstrates how the structures of world time make it possible to use a clock for "reading time off" (*Zeitablesung*), or asking and answering the question, "what time is it?" In the first place, measuring time is made possible by the discovery, as already discussed, of the positions of the sun as it moves across the sky—especially the privileged positions of sunrise, noon, and sunset—as handy for dating events. The sun's visibility and its public accessibility are conditions for Dasein's original discovery, along with nature itself, of a natural clock. In this context, Heidegger mentions not only the sundial but also the "peasant's clock," which requires nothing more than the sun and one's own shadow. As the day progresses, our shadow diminishes until the sun is directly overhead, when it changes direction and grows longer until daylight fades. Prior to the invention of the most rudimentary sundials, one's shadow is available for public time reckoning and even measuring units of time with that shadow's length. Thus, measuring time originally depends on at least two things that are missing from Bergson's account: Dasein's discovery of the sun's regular passage and the changing length and orientation of the shadows it casts, according to its *thrownness* into a world illuminated by the sun, and its dating of events in a way that is publicly accessible, according to its *being-with-others*.

In addition to the discovery of the natural clock provided by the sun, measuring time depends on a certain kind of *discourse* that accompanies looking at the clock. According to Heidegger, "looking at the clock and orienting oneself *according to time* is essentially a *now-saying*," which is nothing other than the "articulation of a *making present* that temporalizes itself in unity with an awaiting that retains" (BT 396/416). Reading time off the clock involves not just looking at it but also *saying*, explicitly or implicitly, "*now* it is one o'clock," "*now* it is one-fifteen," and so on. The clock enables more precise dating, linking events to numbers representing hours and minutes, defining any arbitrary now-point as a particular time of day. Like all dating, using a clock expresses the making present that awaits and retains, and it does so by *fixing the "now"* and providing a standard for measuring the interval between now-points.

Heidegger notes that time measurement has the peculiar characteristic of making something present that is never encountered as objectively present in the world. Prior to the particular case of time measurement, measurement in general has the temporal meaning of making something present together with a standard and defining it in relation to that standard. As Heidegger says, "Measuring is constituted temporally when a present standard is made present in a present length" (BT 396/417). For example, we hold a measuring tape up to a piece of furniture to determine its width, and the units represented by marks on the measuring tape provide a standard that is universal, "present [*vorhanden*] in its constancy at every time for everyone" (BT 396/417). This is simple enough when it is a matter of measuring objects, but the case is more complicated with intervals of time for obvious reasons: an hour or a minute cannot be "held up" to whatever is to be measured because it is not objectively present. A clock provides the necessary standard for measuring time by translating the regular, recurring passage of the sun into equal intervals, but only by virtue of "our eminent making present" (BT 387/417). Insofar as clocks, like the sun, are objects that are accessible to others, "in *measuring time*, time gets *made public*" (BT 387/417). Clocks thereby enable us to represent time "universally" as an "*objectively present multiplicity of nows*" (BT 387/417). According to Heidegger, it is as if we said, when looking at a clock, "now, and now, and now, and now . . ." in such a way that absent "nows" are brought into the present. When reading time off the clock—and this is true with instruments used for more precise time measurement—we take a manifold of now-points that are not actually present and bring them into the present as positions indicated by the hands or as numbers. Thus, answering the question "what time is it?" involves saying "now" in such a way that temporality itself becomes something objectively present.

We can see how this analysis opposes Bergson's view that time measurement is essentially the replacement of duration, a qualitative multiplicity, with space, a quantitative one. Without mentioning Bergson by name here, Heidegger elaborates:

> The time made public in our measurement of it by no means turns into space because we date it in terms of spatial relations of measurement. Nor is what is existentially and ontologically essential in time *measurement* to be sought in the fact that dated "time" is determined numerically in terms of *spatial* distances and changes in the *location* of some spatial thing. Rather, what is ontologically decisive lies in the specific *making present* that makes measurement possible. Dating in terms of what is present "spatially" is so far from a spatialization of time that this supposed spatialization signifies nothing other than that a being that is present [*vorhandenen Seienden*] for everyone in every now is made present [*Gegenwärtigen*] in its own presence [*Anwesenheit*]." (BT 397/418)

This is a key passage for understanding Heidegger's critique of Bergson because it reveals precisely what Heidegger thinks Bergson misunderstands by interpreting time as a spatial representation of duration. In his account of the distinction between pure duration and time as a homogeneous medium in *Time and Free Will*, Bergson argues that the measurement of time is only possible through the translation of duration into a homogeneous medium analogous to space. Heidegger's response is that what is essential to the measurement of time is not "externalization" or "spatialization," but making-present. Space plays a minor role because "nows" are made present with the help of the positions of the sun or the hands of a clock, but Heidegger's view is that measurement as such is a mode of *temporalization*, and it is in these terms that the measurement of time should be understood.

Heidegger's analysis of time measurement is a crucial moment in his engagement with Bergson because it gets to the root of his concerns with Bergson's interpretation of the difference between ordinary time and originary temporality. In the introduction to *Being and Time*, Heidegger declares his opposition to "Bergson's thesis that time understood in the common way is really space" (BT 17/18). According to Bergson, to grasp duration in its "original purity," time has to be freed from all spatial representation—that is, we have to break the habit of thinking about time in terms of space. On Heidegger's account, measured time is not a spatialization of pure duration, but a temporalization dominated by making-present. In circumspect taking care, Dasein temporalizes in such a way that it makes past and future "nows" present with regard to a universal standard. As soon as it is measured, the time taken care of is made present and public as a multiplicity of "nows" occurring at regular intervals. Thus, to distinguish originary temporality from measured time, it needs to be freed from not spatiality, but *presence*.

Heidegger's analysis of time measurement thus clarifies his opposition to Bergson, and furthermore it shows how Bergson's understanding of the relationship between time and space may suffer from his neglect of one basic ontological structure, namely being-with-others. In *Time and Free Will*, Bergson draws a fundamental distinction between the impersonality of time as a homogenous medium and the private or personal character of pure duration. He appears to interpret the publicness and accessibility of measured time as characteristics of space, as if only spatiality were public. However, for Heidegger, the making-present of time in measurement is also a *making-public*. While time is indeed made public by means of things in the environment that move through space (the sun, the hands of a clock), that making-public is grounded in being-with-others. Dasein is not an isolated subject, but shares the world with others, even when it is not encountering particular others. As such, everyday Dasein interprets things in a public way, including its own temporality. For Heidegger, time measurement is not so much a spatialization of time as a making-present of "nows" by means of a publicly available standard, thus a making-public of temporality. Likewise, our ordinary understanding of time is not a spatial representation of the flow of time, but a public interpretation of originary temporality.

If the time that is measured is not, as Bergson argues, a spatial representation, then what about our tendency to think and speak about time in spatial terms, most notably when we talk about things occurring "in time"? In the introduction to *Being and Time*, Heidegger notes that time has traditionally performed the ontological function of distinguishing between two regions of being: the temporal and the non-temporal, things "within time," including "natural processes and historical events," and things "outside of time," such as "spatial and numerical relationships" (BT 18/18). Now the origin of this distinction and the meaning of "within-timeness" or "intratemporality" (*Innerzeitigkeit*) can be understood in terms of in the temporality of Dasein, particularly world time. Heidegger maintains, parallel to the distinction between Dasein's "worldly" (*weltlich*) constitution and the "innerworldly" (*innerweltlich*) character of other beings, that only Dasein is "temporal" (*zeitlich*), strictly speaking, and beings other than Dasein are merely in time, or "intratemporal" (*innerzeitig*). Such beings do not project themselves toward their end, do not come back to themselves from the future, and do not seize upon what they have been in the light of their potentiality-for-being. Beings other than Dasein do not temporalize ecstatically, but instead they are encountered by Dasein in the light of the temporality of circumspect taking care. As innerworldly, these beings show up "in" world time, a time that "has *the same* transcendence as the world" (BT 398/419). Because temporality expresses itself as world time, Dasein interprets everything "in the world" as occurring "in time." Ultimately, for Heidegger, "in time" is not a spatial metaphor, as Bergson assumes, but rather an expression of the "genuine phenomenon" (BT 318/334) of intratemporality—genuine insofar as it not simply the result of confusion, but an indication of how things are encountered within the world.

Heidegger's account of the link between being within time and being innerworldly makes it possible to resolve the question about the objectivity or subjectivity of time. If world time is transcendent in the phenomenological sense, would that not make it objective? Or is world time, as an offspring of originary temporality, subjective? Heidegger's answer is that world time is neither objective nor subjective; more precisely, "*World time is 'more objective' than any possible object*" yet "*world time is 'more subjective' than any possible subject*" (BT 399/419).[29] How so? On the one hand, world time is never something objectively present and never something we encounter as such in the world. Yet it is "more objective" than objects because it shows up in the sky with the passage of the sun, and for that reason, contrary to Kant's theory of time as the form of inner sense, it "is found *just as directly* in what is physical as in what is psychical" (BT 399/419). World time may not be objective in the traditional sense, but it is always already given along with the world; thus, it is an objectification of temporality. On the other hand, because world time is public, "available for everyone," it is not merely subjective. Yet it is "more subjective" than subjects because it is a condition for the possibility of "*the factically existing self*" (BT 399/419). That is, care, which is the "being of the subject" from an existential-ontological standpoint, is made possible as circumspect taking care by world time. Thus, world time is the condition for the possibility of Dasein temporalizing itself as an abiding subject of experiences. Heidegger concludes somewhat enigmatically: " 'Time' is present neither in the 'subject' nor in the 'object,' neither 'inside' nor 'outside,' and it 'is' '*prior*' to every subjectivity and objectivity, because it presents the condition of the very possibility of this 'prior' " (BT 399/419).[30] Although we may be tempted to read Heidegger as following Kant and Husserl in conceiving time as a subjective feature of human experience, here he maintains that time, and *a fortiori* temporality, is ontologically prior to the very distinction between subject and object.[31] He seeks to avoid "volatilizing" time (as Bergson is often interpreted as doing) or "reifying" it (as Bergson claims that philosophers and scientists tend to do), as well as "vacillating insecurely between" these options (BT 399/420).

6. The Origin of the Concept of Time

Heidegger's account of world time has provided a bridge from ecstatic temporality to time as it is ordinarily understood, enabling him to show how the concept of time is derived from temporality. Whereas Bergson regards the ordinary concept of time as an abstract, spatial representation of concrete, lived duration, Heidegger views it as an interpretation of temporality as objectively present. If time is an "illegitimate concept" (TFW 98/66) and a "symbolical substitute" (TFW 124/82) for Bergson, on the contrary, Heidegger aims to establish its "rightful autonomy" (BT 17/18). Using Aristotle's definition of time as a point of departure, Heidegger argues that the concept of time "makes time thematic in the way that it shows itself in circumspect taking care" (BT 400–1/421). This "thematic" or conceptual understanding of time is not, as Bergson suggests, duration in "the illusory form of a homogeneous medium" (TFW 110/73). Instead, it is an interpretation of Dasein's temporality as a "succession of nows" that is "something somehow objectively present" (BT 402/423). This "something" is what is called time. Although the ecstatic unity of temporality is obscured by the concept of time, Heidegger insists that it is no mere illusion:

> The vulgar characterization of time as an endless, irreversible succession of nows passing away arises from the temporality of entangled Dasein. *The vulgar representation of time has its natural justification.* It belongs to the everyday kind of being of Dasein and to the understanding of being initially dominant. (BT 405/426)

Heidegger is concerned not just to criticize the concept of time or dispel the confusion surrounding it, but rather to show how it originates in the temporality of Dasein and bears the imprint of its origin. To do this, he examines how the ordinary concept of time is a reflection of Dasein's falling prey to or entanglement in the "they," on the one hand, and the traditional understanding of being as presence, on the other. That is, he traces the concept of time to both the phenomenon of inauthentic temporality and the ontology of the ancient Greeks.

World time, on Heidegger's analysis, is an expression of the "making present that awaits and retains," prior to any conceptual understanding of time. World time is not (yet) conceived as a succession of "nows," but it is expressed by saying "now," "then," and "on that former occasion." The "now" is not (yet) a "point of time"; rather, as a phenomenon of world time, it is datable, spanned, public, and significant. How does this phenomenon of world time—seemingly very different from, if not contradictory to, the "succession of nows"—come to be interpreted this way? How do the ordinary understanding of time and the traditional concept of it arise from Dasein's taking care of time?

Although we constantly deal with time and regulate our daily lives according to it, when we turn our theoretical gaze toward it, the structure of world time is obscured. What shows up when we stop taking care of time circumspectly and make it a theme of investigation is, in Heidegger's terms, the "time of nature" (*Naturzeit*). World time and the time of nature are not two different things, but two ways temporality appears, the first for circumspect taking care, and the second for the "natural understanding." In the latter, illustrated by Aristotle's thought, time is grasped as a part of nature and defined in terms of objectively present things that change and move in regular, predictable ways. As Heidegger says in the introduction to *Being and Time*, he regards Aristotle's definition of time as "that which is counted in the motion encountered in the horizon of the earlier and the later" (BT 400/421) as the first formulation of natural time. To clarify the relation between this concept of time and the temporality of Dasein, Heidegger explains what makes Aristotle's definition of time a *natural* interpretation of temporality.

The ambiguity of "natural" is significant here. In one sense, Aristotle's way of understanding time is natural insofar as it deals with a common phenomenon and expresses *something self-evident* about it. This captures part of Heidegger's meaning, but the self-evidence of a proposition is problematic if what makes it self-evident is that the experience grounding it has been forgotten or covered over.[32] What is puzzling about time arises not only from its being so familiar we tend to overlook it, but also from the particular ontological perspective adopted by Aristotle and handed down to us. His definition of time, the guide for "all subsequent discussion of the concept of time" (BT 400/421), is natural in another sense. Aristotle's interpretation is based on a preconception of time as something present in nature along with all things capable of motion and rest. As an accessory to the experience of motion, time is seen as *a feature of the natural world*. Aristotle's theory is oriented toward what is objectively present, for it rests upon a "natural understanding of being" (BT 400/421) as presence (*parousia*). Thus, Aristotle's definition is "natural" in two senses: it is drawn from a common phenomenon, giving it the appearance of self-evidence, and it is determined by the ontological presupposition that being means *presence*.[33]

On the basis of this natural understanding of being, it is present time—the now—that is primary. According to Aristotle's definition, "Time is 'what is counted' . . . in the making present of the *moving* pointer (or shadow)" (BT 401/422).[34] Because Dasein always says "now" when reckoning with time and implicitly says "now, and now, and now . . ." when reading time off the clock, Heidegger explains, "What is counted are the nows" (BT 401/422). For this reason, the temporality that shows itself in the experience of using clocks deserves to be called "now-time" (*Jetztzeit*). Aristotle's account of time, like the discovery of natural clocks and the development of artificial ones, thus arises from the temporal mode of the "making present that awaits and retains." The *traditional concept* of time is a theoretical formulation of the *ordinary understanding* of time as something in which beings are encountered as either "now," "not yet (present)," or "no longer (present)." In this mode, "time shows itself for the vulgar understanding as a succession of constantly 'present' nows that pass away and arrive at the same time. Time is understood as a sequence, as the 'flux' of nows, as the 'course of time'" (BT 401/422). In this respect, "now-time" is a faithful description of both the traditional concept of time handed down by Aristotle and the common understanding of it.

The time of nature, or now-time, is an interpretation of world time as something objectively present, and its structure reflects its provenance. According to Heidegger, the concept of time "covers over" (*verdeckt*) the structure of world time and "levels down" (*nivelliert*) ecstatic temporality. When Heidegger claims that the concept of time covers over world time, he means that it hides some of the essential characteristics of world time from view. The primary thing that distinguishes now-time as a concept from the complete structure of world time is this: "In the vulgar interpretation of time as a succession of nows, both datability and significance are *lacking*" (BT 401/422). As soon as time is understood as a "succession of nows," the relational structures that are essential to world time disappear. The aspect of the "now" through which it is always referred to events, the "now, that . . . ," is hidden. Likewise, the appropriateness of each now for certain things and its inappropriateness for others is concealed. In the "pure sequence" of time, the nows are identical, distinguished only by their position in relation to one another or to the present. Their relations of reference to what Dasein takes care of are thus obscured: "The nows are cut off from these relations, so to speak, and as thus cut off, they simply range themselves along after one another so as to constitute the succession" (BT 401/422). With this, not only is world time covered over, but ecstatic temporality is also leveled down. This means that the ecstasies no longer appear as the movements of coming toward, going back, and taking hold that constitute the unity of Dasein's temporality. Instead, temporality appears to consist of a multitude of nows that come out of the future, pass through the present, and recede into the past. For the ordinary understanding, the now is primary, and while only the present now is real, there are no singular or exceptional nows, only an identical series. Temporality as a whole is made present, and Dasein understands the past and the future as "closed off," as regions accessible only through the present in memory (or retention) and expectation (or awaiting).

In "The Concept of Time," Heidegger spoke in Bergsonian terms of the "homogenization" of time into now-points, "an assimilation of time to space, to Presence pure and simple" (CT 18/121). With this derivation of the concept of time from world time, however, he rejects the notion that what is at issue is a representation of time in spatial terms as Bergson's error and instead focuses on the temporal mode of making-present. The ordinary concept of time, Heidegger argues, is not a spatialization of temporality, but an interpretation of it as something that is encountered together with innerworldly things. In this sense, "[t]he succession of nows is interpreted as something somehow objectively present" (BT 402/423). For Heidegger, the basic features of time as it is ordinarily understood are all derived from the characteristics of world time and are necessary concealments of temporality.

To demonstrate that now-time, or the time of nature, is an interpretation of world time, Heidegger shows how its essential characteristics are derived from those of world time and, ultimately, originary temporality. First, the *continuity* of time can be traced back to the "spannedness" of world time, which refers to how Dasein "stretches along" from day to day. Although there may be "gaps" of time unaccounted for in memory, we ordinarily understand time as a succession that "is uninterrupted and has no gaps" (BT 402/423). This is only the case, according to Heidegger, thanks to the unity of ecstatic temporality, which enables Dasein to make things present not just in now-points, but also in spans of time. The continuity of time remains puzzling, however, particularly how time can be both continuous and divided into "nows." This is not because we confuse duration with space, as Bergson argues, but because we confuse the temporality of Dasein with the presence of mere things. The continuity of something objectively present, something that constantly remains what it is, serves as the model for time, which is "always there" with whatever is encountered in it.

Second, the *infinitude* of time, which contrasts most sharply with originary temporality, can be traced back to the "publicness" of world time. Dasein's temporality is finite, according to Heidegger, insofar as Dasein is being-toward-death. Not only does Dasein "come to an end," but also Dasein *is* its end in the sense that death is its most certain but indefinite possibility. In being-toward-death, Dasein "exists finitely," and in anticipatory resoluteness, temporality "reveals itself *as finite*" (BT 315/329–30). With Dasein's flight from death, however, this finitude is forgotten and covered over by a concept of time as infinite. When time is understood as a "succession of nows," every one coming before some and after others, we cannot conceive of a beginning or end to time: "In this *way of thinking* time through to the end, one *must* always think more time; from this one concludes that time *is* infinite" (BT 403/424). What is the reason for the necessity of thinking more time? This is a way of fleeing death by "looking away from finitude" (BT 403/424). Every life has an end, but public time is "endless 'in both directions'" (BT 403/424). I know that when I die, my time has come to an end, but when *one dies*, according to the public interpretation, "[t]ime goes on. . . . One only knows public time that, leveled down, belongs to everyone, and that means to no one" (BT 404/425). Thus, time is infinite insofar as it is public, the time of the "they," which conceals the finite temporality of Dasein.

Two more characteristics of time can also be explained in terms of Dasein's flight from death and concealment of ecstatic temporality. The third is the *passing* of time, the idea that time is fleeting, which is also a characteristic of Dasein's knowledge of death. When we speak of time as passing, we mean that the succession of nows cannot be stopped, and for Heidegger this implies a wish that it could be stopped, that we could escape death. The passing of time and the end of life are dual inevitabilities, and neither time's unceasing flow nor the approach of death can be halted. The fact that we emphasize time's "passing away" shows that our concept of time still dimly reflects the finitude of originary temporality: "Dasein ultimately understands more than it would like to admit, that is, the *temporality* in which world time temporalizes itself is *not completely closed off* despite all covering over" (BT 404/425). Dasein's flight from death also sheds light on a fourth characteristic of time as it is ordinarily understood, its *irreversibility*. If the succession of nows is without beginning or end, then what would prevent it from flowing in the opposite direction? For Heidegger, the irreversibility of time reflects, in distorted form, the primacy of the future in the temporality of Dasein. The impossibility of reversing the direction of time, like the impossibility of stopping it, is a public interpretation of the movement of ecstatic temporality. As futural, Dasein always already *is* its end, but this coming toward itself is usually understood as a unidirectional movement of time toward the past through the present. With the idea of irreversibility, once again originary temporality shows up in the way time is ordinarily understood, but in a distorted or disguised form.

Heidegger's strategy for describing how the concept of time is derived from ecstatic temporality is not simply to contrast common representations of time with immediate experience, as Bergson does in *Time and Free Will*. In fact, throughout this discussion, Heidegger works to distance his interpretation of temporality from the notion of time as a "sequence of experiences." Instead, his strategy has been to show how originary temporality expresses itself in Dasein's circumspect taking care as world time, and then how world time undergoes a "covering over that levels down" (BT 405/426) with the interpretation that is most "natural" to everyday Dasein. Thus, rather than attempting to reverse or overturn the traditional concept of time, Heidegger aims to show how it both conceals and reveals the structure of originary temporality from which it arises.

The discussion of now-time serves two purposes with respect to Heidegger's attempt to clarify the relationship between temporality and the concept of time. On the one hand, it shows that this concept is not simply illegitimate, as Bergson claims, but "arises from the temporality of entangled Dasein" and "*has its natural justification*" (BT 405/426). For Heidegger, within the limits of everydayness and in the context of measuring and reckoning with time, the ordinary way of understanding time is justified. If the beings that Dasein is "together with" may be said to be "in the world," then they may rightly be said to be "in time," or in the opening maintained by temporality. What is unjustified, in Heidegger's view, is the notion that this concept of time is the only true one. Thus, on the other hand, the discussion of now-time and its characteristics reveals not only how it is justified, but also why it is not the *primordial phenomenon of time*. Everyday Dasein understands the beings it encounters in terms of objective presence, including things (such as death) that cannot be present. The concept of time as a succession of nows is a making-present of temporality, whereas temporality is the *condition of presence*. Temporality cannot be wholly present or given as such without being essentially altered—"leveled down" and "covered over," in Heidegger's words. Dasein understands temporality, for temporality is the meaning of its being, but its natural understanding, defined by the privilege of presence, conceals the origin of time. Heidegger finally articulates the most crucial difference between time and temporality, in his view—the difference that explains why he refers to *both* the traditional concept and the ordinary way of understanding time as "now-time":

> Ecstatic and horizonal temporality temporalizes itself *primarily* out of the *future*. However, the vulgar understanding of time sees the fundamental phenomenon of time in the *now*, and indeed in the sheer now, cut off in its complete structure, that is called the "present." (BT 405/426–7)

The concept of time is thus essentially determined by the *primacy of the now*, as is evident in Aristotle's *Physics*. His definition of time as "something counted with respect to motion" is problematic not because it treats time as quantitative, but because it assumes that what is counted, namely a succession of nows, is what defines time as such. On the level of the ordinary understanding, the now is viewed as fundamental because circumspect taking care, the mode in which Dasein reckons with time, is, in terms of time, a "making present that awaits and retains." The now is also viewed as fundamental when time is conceptually formulated on the basis of the ontological presuppositions—which could be considered prejudices—that being is equal to presence, whatever exists must be objectively present, and therefore the only time that truly exists is the time of the present. For Heidegger, now-time springs from the temporality of Dasein and deserves to be considered as a genuine phenomenon, not an illusion, as long as we realize that it is a public, inauthentic substitute for a more primordial phenomenon.

We are now in a good position to take stock of how Heidegger's path from temporality back to the concept of time has led away from Bergson's thought. By deducing the characteristics of now-time from ecstatic temporality, using the structure of world time as a bridge between them, Heidegger rejects the notion that time is a product of spatial thinking. What we call time is, on his account, the way that ecstatic temporality gets interpreted on the basis of the "natural" understanding of being and the privilege of presence. As such, being "in time" means being objectively present or existing in nature—both referring back to the beings that Dasein encounters in the world. This is not, as Bergson suggests, a crude spatial representation, but an ontological feature of innerworldly beings, which are "intratemporal" as opposed to being "temporal" the way that Dasein is. For Heidegger, the "in" of being in time is, like that of being-in-the-world, not a spatial relation but an existential one. Bergson is quick to conclude that time is not just a derivative concept, but also an illegitimate one. As a consequence, Heidegger thinks, he misses the structure of world time and, more importantly, ecstatic temporality. While Bergson recognizes that originary temporality is concealed by the way time is ordinarily understood, he mistakes duration, which is only a feature of world time, for the primordial phenomenon. It is not enough to distinguish duration, as lived time, from spatial representations of time. As long as duration is conceived as a succession of nows, we remain at the level of the ordinary understanding, which regards time as something objectively present. Duration is no different from time as a measurable quantity with regard to the primacy of the present, so duration is still a form of now-time. Bergson fails to see this, on Heidegger's interpretation, because he misdiagnoses the problem with our ordinary concept of time, which is not a spatialization of duration, but a making-present of temporality.

7. Temporality and Spatiality

While Heidegger is adamant that the problem with the traditional concept of time is not the intrusion of *space* but the privilege of *presence*, he acknowledges our tendency to speak and think about non-spatial things in spatial terms. Even as he argues in *Being and Time* that the spatiality of Dasein is grounded in its temporality, he notes that "both the self-interpretation of Dasein and the stock of meaning so pervasive in language in general are thoroughly dominated by 'spatial representations'" (BT 351/369).[35] As Bergson points out, this is evident in many common ways of thinking about time: the image of time as a line and its moments as geometrical points, the metaphor of duration as a length of time, and the familiar trope of time as a container in which everything happens, for example. Heidegger does not believe that Bergson's approach of distinguishing duration from space is sufficient, because it leaves intact the privilege of presence. Nevertheless, Heidegger's interpretation of temporality reveals a stringent avoidance of spatial metaphors commonly used for thinking about time. He does not make the critique of spatial thinking his overt aim and in fact attacks Bergson for doing so. However, a close examination shows that Heidegger describes originary temporality and its offspring, world time, in decidedly non-spatial terms.[36] This is no accident, for even in the face of his claims that Bergson wrongly identifies ordinary time with space, merely reverses Aristotle's way of thinking about time, and fails to reach originary temporality, Heidegger himself seeks to disentangle temporality from spatiality. This is evident not only from the conspicuous absence of spatial language in his descriptions of world time, but also from his efforts to demonstrate that Dasein's spatiality springs from originary temporality. It is also evident from his interpretation of the world, in contrast to Descartes's conception of it as *res extensa*, as a phenomenon that is not primarily spatial.[37]

First, in Heidegger's discussion of world time, the bridge between originary temporality and the ordinary concept of time, he carefully avoids common spatial metaphors for time. He begins with a set of expressions lacking any obvious spatial connotation—"now," "then," and "on that former occasion"—to indicate the relational structure of datability. He avoids the kind of mistakes highlighted by Bergson in *Time and Free Will* by distinguishing the now of world time from a *point*. Instead, he treats the now as something that is spoken, a linguistic expression of making-present. In taking care of things encountered in the world, saying "now" expresses something always already "there" for Dasein but never objectively present, namely temporality. Heidegger explains that it is thanks to the natural understanding of being that a "now, that . . ." comes to be interpreted as a "point in time" (BT 388/407), but as such this metaphor has an obviousness that is rarely questioned. When we recognize that the common understanding of the now as a point implies objective presence, its ontological inappropriateness becomes clear. In his discussion of world time, Heidegger assiduously avoids characterizing the now as a point, and in this way he breaks one of our most ingrained habits of representing time in terms of space.

Heidegger also avoids any mention of *length* in his account of world time, substituting the terms "span" and "stretch" to characterize temporal extension. As with datability, he approaches the structure of spannedness through adverbial expressions such as "until" and "while," leaving aside equally common ways of speaking, including all talk of "longer" or "shorter" periods of time. Avoiding not only spatial language, but the Bergsonian question of whether duration is something quantitative or qualitative, Heidegger characterizes duration as another feature of Dasein's self-interpretation (BT 390/409). We think of time as enduring because we ourselves are "stretched along" from birth to death, as he emphasizes in his discussion of temporality and historicity. Not only a lifetime, but also a day and even a moment appear, in the light of Dasein's making-present, as having length—even if that length vanishes into the now. But again, this interpretation treats the now as something objectively present. Instead of talking about the length of the now, Heidegger speaks of it as always having a span, at least in the context of world time. This is his attempt to think about temporality as extended in a way that is ontologically prior to both extension in space and duration in time.[38]

Heidegger's discussion of the publicness and significance of time not only avoids spatial metaphors but also suspends spatial thinking by focusing on the worldliness of world time. Time is public, on his account, not because it exists "out there" in the world, but for the reason that "people can say '*now*' together" (BT 391/411). This is possible by means of things encountered in the world—first the sun and later clocks—that enable Dasein to date events with others. Contrary to Bergson's argument that duration cannot be measured without being projected into space, Heidegger characterizes the measurement of time as an exemplary form of both making-present and making-public. What is crucial, in his view, is not that world time is spatial or external, but that it is publicly accessible, a phenomenon of being-with-others. This time is not only public, but also significant because it is understood as appropriate for some activities but not for others. Thus, world time is never homogeneous, but is "time for" (*Zeit zu*) something. This is the case because taking care of time is determined by the relational structure of the world, in which things show up for Dasein as appropriate for some uses but not for others. The "in-order-to" (*um-zu*) that links innerworldly things to one another in a referential totality also pertains to the time that is datable, spanned, and public. The time taken care of deserves to be called "world time" because it "essentially has the character of world" (BT 394/414). This provides the key to understanding the most pervasive spatial representation of all: that things happen "in time." Rather than avoiding this language, Heidegger asks about its existential meaning, the meaning of *intratemporal* (*innerzeitig*). His answer is that being in time means being *innerworldly* (*innerweltlich*), or encountered by Dasein within the world. In his words, "world time has *the same* transcendence as the world," so that when Dasein is together with innerworldly beings, it "understands those beings as circumspectly encountered 'in time' " (BT 398/419). Conversely, beings that are never encountered within the world—such as geometrical axioms, logical relations, and anything traditionally considered to be absolutely invariable—are understood as outside of time or extratemporal (*außerzeitig*). Without directly confronting Bergson's criticism of the idea of time as a homogenous medium or container in which events unfold, Heidegger deflates spatiality by giving the "in" a non-spatial sense.

Heidegger's avoidance of spatial language and his non-spatial interpretation of "in time" do not provide conclusive evidence that he, like Bergson, is interested in disentangling temporality and spatiality. But there is more evidence in Heidegger's account of how Dasein's spatiality arises from temporality. In his discussion of the measurement of time, where he denies that world time is a mixture of time and space, he reiterates that temporality is "what primordially makes the opening up of space possible" (BT 397/417). His view is not that temporality is something completely distinct from space, but that "the specific spatiality of Dasein must be grounded in temporality" (BT 349/367). According to the ordinary understanding, space and time are two basic ways that human beings, like everything in nature, are determined, with neither being reducible to the other. However, from an existential standpoint, the spatiality and temporality of Dasein cannot be the same as that of something objectively present "in space and also in time" (BT 349/367). The temporality of Dasein is, according to Heidegger, the origin of the notion that things occur "in time" (cf. BT 392/412). Furthermore, Dasein's temporality, as the meaning of care, grounds all of its existential structures, including spatiality. Thus, in opposition to the view of space and time as independent and "equiprimordial," Heidegger argues that Dasein's spatiality is grounded in temporality. This is not Bergson's way of disentangling time from space, yet it shows how important it is for Heidegger, at least at this stage, that originary temporality is not just distinct from spatiality, but also ontologically prior to it.[39] But what is spatiality, and how, exactly, is it grounded in temporality?

Heidegger's account of spatiality (*Räumlichkeit*) in division one of *Being and Time* begins with the handiness of things in Dasein's surrounding world and their "nearness" (BT 99/102ff). Prior to the manifold of pure geometrical or physical space, he argues, Dasein encounters things as nearby or out of reach, right here or over there, in their place or missing, and so on. As such, circumspect taking care discovers useful things that belong to various *regions* prior to the concept of space as a three-dimensional container inhabited by objects. The spatiality of such innerworldly things depends on the "de-distancing" (*Ent-fernung*) and "directionality" (*Ausrichtung*) of Dasein, which open up these regions. These terms refer to the way everyday Dasein discovers distance by bringing things near and makes it possible to locate them by referring them to their places. Things appear "in space" thanks to Dasein's existential spatiality, which is nothing but a way of "letting innerworldly beings be encountered," that is to say, "giving space" (*Raum-geben*) or "making room" (*Einräumen*) for beings (BT 108/111). Thus, like the temporality that "gives time," Dasein's spatiality is a condition for the possibility of things appearing in space. In Heidegger's view, though, this is not, as Kant and even Bergson maintain, a subjective condition of experience. Instead, existential spatiality is an ontological structure of being-in-the-world prior to the subject-object distinction.

It is with this interpretation of spatiality in the background that Heidegger later claims: "Only on the basis of ecstatic and horizonal temporality is it possible for Dasein to break into space" (BT 351/369). The existential spatiality that Heidegger has described in terms of the de-distancing (or bringing near) and directionality of being-in-the-world is primarily a matter of making-room. As the way that Dasein discovers the regions in which handy things belong, this mode of orientation is ontologically prior to having a representation or concept of space. The discovery of things as belonging here or there depends on the disclosure of the world as a referential totality, which in turn depends on ecstatic temporality. More precisely, "[t]he self-directive discovering of a region is grounded in an ecstatically retentive awaiting," and "[b]ringing-near and the estimating and measuring of distances . . . are grounded in a making-present that belongs to the unity of temporality" (BT 351/368–9). According to Heidegger, space is first "opened up" thanks to the ecstatic openness of Dasein, and things appear present in space on the basis of "the making present that awaits and retains," a mode of temporality. Consequently, just as the "now" of circumspect taking care is not simply a "point in time," likewise "the here of [Dasein's] actual factical location or situation never signifies a position in space, but the leeway [*Spielraum*] of the range of the totality of useful things taken care of nearby" (BT 351/369). This leeway is made possible by the way Dasein temporalizes in the mode of taking care.

Although Heidegger goes into less detail about the emergence of space from Dasein's spatiality than he does about how the time of nature springs from world time, he does offer a few key points. To begin with, he stresses the difference between Dasein's spatiality and that of things objectively present in space. By making room, or bringing near, Dasein is spatial, yet "Dasein is never objectively present in space, not even initially" (BT 350/368). Why, then, do we end up thinking of not only the objects we encounter in the world but also ourselves as "extended things" that "take up space"? Heidegger explains this on the basis of inauthentic temporality. On the one hand, Dasein's absorption in things taken care of, the things nearby, can be traced to a making present that "loses itself in itself, and forgets the over there" (BT 351/369). On the other hand, this making present is equally a forgetting of the existential spatiality prior to the presence of things in space. Forgetting how we ourselves open up various regions in accordance with our concerns, we fall prey to a public interpretation of space. The ordinary concept of space as a manifold of points subtending external objects—in Bergson's terms, an empty, homogenous medium—is, like the concept of time, a "natural" understanding of space as something objectively present. Unifying all the regions and all the places into a totality that is characterized by constant presence, we develop the concept of "space in general," in which extended things, including ourselves, are located.

So when Heidegger attempts to disentangle time from space, it is by no means to suggest that temporality is completely separate from spatiality. He does, however, give temporality a certain ontological priority, which should not be confused with the way Kant understands the priority of time. Although Heidegger claims that "Kant is the first and only one who traversed a stretch of the path toward investigating the dimension of temporality [*Temporalität*]" (BT 22/23), he denies that time is prior to space in the way that Kant maintains in the Transcendental Aesthetic. Thus, the ontological priority of temporality for Heidegger should not be confused with the supposed priority of time, as the form of inner sense and condition of all objects of experience, over space, as the form of outer sense and condition of external objects only. According to him, the Kantian view only vaguely hints at the way that both time and space as ordinarily understood arise from temporality. Moreover, the existential-ontological priority of temporality does not preclude the priority of space in our thinking. In a concession to Bergson's view of why we commonly confuse time with space, Heidegger acknowledges the predominance of "spatial representations" in the "self-interpretation of Dasein and the stock of meaning so pervasive in language in general" (BT 351/369). However, for Heidegger, this "priority of the spatial" is a result of the way that making-present dominates our average understanding of ourselves and the world. Thus, what Bergson diagnoses as spatial thinking is indeed problematic for Heidegger, but he insists that it is rooted in a deeper ontological problem.

8. Temporality and Selfhood

Heidegger's remarks about Bergson in *Being and Time* highlight his concerns with the way Bergson seeks originary temporality by distinguishing duration from space, but Heidegger says nothing about how Bergson attempts to rethink the ego or self (*moi*) in terms of duration. As a result, the similarity between Heidegger's interpretation of "selfhood" (*Selbstheit*) and Bergson's theory of the "two aspects of the self" in *Time and Free Will* has generally been ignored.[40] In much the same way that Bergson contrasts the "fundamental" self as unfolding in duration with the "superficial" self projected into space, Heidegger distinguishes between authentic and inauthentic selfhood as different ways of temporalizing. To begin with, both philosophers offer accounts of how the self is shaped by "social life" or "publicness" and how our ordinary understanding of the self raises obstacles to freedom. Both also develop accounts of selfhood as a temporal phenomenon, calling into question the unity and permanence of the ego. Like Bergson, Heidegger uses Kant's philosophy as a point of departure and criticizes his conception of subjectivity. Finally, while Bergson argues that a free act is performed by the fundamental self and requires us to "get back into duration," Heidegger also interprets freedom in *Being and Time* as a matter of "self-constancy" (*Selbstandigkeit*), which depends on authentic temporality.

Prior to Heidegger's temporal interpretation of the self in division two of *Being and Time*, we can already see the similarity between his view and Bergson's in division one, where he discusses two obstacles to Dasein's self-understanding. First, Heidegger argues that average, everyday Dasein understands itself on the basis of non-human things, "*those* beings, and their being, which it itself is *not* but which it encounters 'within' its world" (BT 59/58). Dasein typically understands itself not as *existing*, in Heidegger's sense of the term—that is, not as a being defined by its possibilities—but as objectively present. As such, we tend to think of ourselves as things with properties, but as things that differ from others by virtue of some special property, such as the capacity for reason, language, or consciousness. This reifying or objectifying tendency of the ordinary understanding is the basis for the traditional interpretation of Dasein as "the 'subject,' the 'self' . . . [that which] maintains itself as an identity through the changes in behavior and experiences . . . something self-same in manifold otherness" (BT 112/114). In short, even when we claim to have made a fundamental distinction between subject and object, consciousness and things, mind and matter, and so on, in Heidegger's view we continue to think of the self in terms of objective presence.

The second obstacle is that the average, everyday self-understanding of Dasein is dictated by others, specifically by "those from whom one mostly does *not* distinguish oneself" (BT 115/118). For Heidegger, knowing oneself and, deeper still, "being a self" (*Selbstsein*) are grounded in "being-with" (*Mitsein*). For average, everyday Dasein, self-knowledge and selfhood are thus characterized by an "inconspicuous domination by others that Dasein as being-with has already taken over unawares" (BT 123/126). Dasein understands itself not only in terms of objective presence, but also in terms of the "they" (*das Man*). Heidegger writes: "The self of everyday Dasein is the *they-self*, which we distinguish from the *authentic self*, that is, the self which has explicitly grasped itself. As the they-self, Dasein is *dispersed* in the they" (BT 125/129). Accordingly, we tend to understand ourselves in terms of what "one does" or "they do," and this is what it means to be inauthentic, for Heidegger. Authentic possibilities are leveled off, "everything that is original is flattened down as something long since known" (BT 123/127), and "[p]ublicness obscures everything and then claims that what has been thus covered over is what is familiar and accessible to everybody" (BT 124/127). Thus, we think of the self as something stable, self-identical, and "constant" not only because we understand ourselves in terms of objective presence, but also because our actions and thoughts are constantly dominated by the "they."

While Heidegger and Bergson employ a different vocabulary, their descriptions of inauthentic and superficial selfhood bear several significant similarities. For one thing, Heidegger portrays average, everyday Dasein as falling away from authentic selfhood and into absorption in worldly concerns. He describes inauthenticity as "a distinctive kind of being-in-the-world which is completely taken in by the world and the Dasein-with of others in the they . . . the kind of being of Dasein closest to it and in which it mostly maintains itself" (BT 169/176). In Heidegger's view, Dasein's absorption amounts to a "fleeing in the face of itself," a turning away from its authentic potentiality-for-being. Likewise, Bergson describes the loss of the fundamental self as a result of our turning away from the "confused, ever changing, and inexpressible" aspect of conscious life, which "language cannot get hold of . . . without arresting its mobility or . . . making it into public property" (TFW 129/85). For Bergson, too, we are "taken in" by the world and others because "social life is more practically important to us than our inner and individual existence" (TFW 130/86). Heidegger's contention that a "public interpretation" of the self covers up authentic selfhood echoes Bergson's claim that "social life in general and language in particular" cause us to "lose sight of the fundamental self" (TFW 128/85). Both thinkers argue, at least in these early works, that authentic individuality requires an effort to recover the self from conventions of language and thought.

Another significant similarity is the way Heidegger and Bergson both reject traditional accounts of freedom and attempt to rethink it in terms of temporality. On Bergson's account, "[t]o act freely is to recover possession of oneself, and to get back into pure duration" (TFW 231/151). However, freedom is not a property of every conscious decision, but a characteristic of certain acts in which we do not merely choose from the possibilities available to us, but we do so in a way that calls upon a depth or wholeness that for the most part is lacking. That is, freedom is an expression of the person as a whole, or the fundamental self in its capacity to disrupt the "conscious automatism" of the superficial self. For Heidegger, anticipation is the mode of temporality that enables Dasein to face "*the possibility to be itself . . . in the passionate anxious freedom toward death, which is free of the illusions of the they*" (BT 255/266). The anticipation of death is key, for Heidegger, in disclosing Dasein's "authentic *potentiality-of-being-one's-self*" (BT 257/267) and thereby modifying the inauthentic "they-self." On his account, inauthentic Dasein, falling away from itself, does not choose for itself, but is carried along by what "they" say and do, "taken along by the no one, without choice" (BT 258/268). Inauthenticity is "lostness in the they," a condition from which Dasein must be "brought back" to itself. Dasein "becomes *authentic* being-one's-self" by not only choosing, but also "choosing to make this choice" (BT 258/268). Authentic selfhood, as a liberating recovery from the lostness of inauthentic selfhood, thus turns out to be a mode of temporality or way of temporalizing.

Heidegger's attempt to understand selfhood in temporal terms becomes more explicit when he calls authenticity and inauthenticity "fundamental possibilities of existence . . . ontologically grounded in possible temporalizations of temporality" (BT 291/304). As opposed to the constant leveling down of the "they-self," as well as the constancy (*Beständigkeit*) of anything objectively present, Heidegger characterizes authentic selfhood as "self-constancy" (*Selbständigkeit*). He argues that Kant points the way to an understanding of the self as something that is never objectively present, like the Cartesian subject. On the one hand, he praises Kant's attention to the "phenomenal content of saying I" and his recognition of the "impossibility of ontically reducing the I to a 'substance'" (BT 304–5/319–20). On the other hand, he criticizes Kant for defining the "I" as the form of representation that makes all other representations possible, thus something *constantly present* in the manner of Descartes's *res cogitans*. According to Heidegger, although Kant understands the "I think" as a formal structure, a principle of unity rather than a substance, "he still slips back into the same inappropriate ontology of the substantial" (BT 304/319). The Kantian subject is defined by an *a priori* self-sameness or constancy that Heidegger regards as characteristic of objective presence rather than existence. In his view, Kant remains traditional insofar as he overlooks being-in-the-world, as well as the difference between authentic and inauthentic saying "I." Authentic selfhood, for Heidegger, is not a given but an achievement, which must be understood in temporal terms: "*The constancy of the self* in the double sense of constancy and steadfastness is an *authentic* counter-possibility to the unself-constancy of irresolute falling prey. Existentially, the *constancy of the self* means nothing other than anticipatory resoluteness" (BT 308/322). In other words, "self-constancy" is another term for authentic temporality, which is a possible way of being, not a condition for the possibility of saying "I." Although Kant and many of his followers recognize that the self cannot be adequately understood as a substance or thing, they fail to see that even a purely formal "I think" is problematic. For Heidegger, both the constancy and inconstancy of the self must be explained in terms of the ecstatic unity of temporality.

Although Heidegger says nothing in *Being and Time* about Bergson's distinction between two aspects of the self, the fundamental and the superficial, his interpretation of authentic selfhood as self-constancy represents an attempt to think with and against Bergson. Bergson shows in *Time and Free Will* that our conception of the self is impeded by the tendency to objectify or reify it; that this self-understanding is superficial and dictated by social needs; that a deeper, more authentic self can be discovered with effort; and that this self must not be understood as something permanent or abiding—not as a substance or subject—but as something fundamentally temporal. *With* Bergson, Heidegger challenges the assumption that the self or ego is an unchanging foundation for experiences, and he attempts to rethink selfhood in terms of time. However, the distinction between authentic and inauthentic existence, as Heidegger develops it in *Being and Time*, is clearly addressed to more thinkers than Bergson. For instance, we can hear echoes of Kierkegaard in Heidegger's notion of the public and its "leveling down" of individuality, as well as his interpretation of the authentic present as a "moment of vision" (*Augenblick*). Furthermore, Heidegger thinks *against* Bergson by rejecting the notion that what distinguishes the inauthentic self is spatial thinking and instead showing how the privilege of presence leads to the misconception that the self must be objectively present in some way. Thus, Heidegger's target is not just the representation of the self in modern philosophy and psychology, the "mechanical conception" that Bergson attacks, but also the presupposition that permanence of the self is necessary for one to say and think "I am." With respect to selfhood, the confusion he aims to dispel is not between duration and space, but between the temporality of Dasein and objective presence. For Heidegger, authenticity cannot be just a matter of searching beneath the symbolic representation of the self for its underlying reality, for even our understanding of reality is based on the assumption that being means presence.[41]

9. Heidegger's Bergsonism

At the heart of *Being and Time* are Heidegger's claims that temporality is the meaning of Dasein's being and time is the horizon for understanding being. In this regard, two demands of Heidegger's project are clear from the very beginning: to pursue the question of the meaning of being, we must consider the kind of being that is peculiar to the entities that can ask such a question, and to understand this kind of being, we must focus on time. To fulfill these demands, Heidegger begins with an existential analysis of Dasein and deepens that analysis with a temporal interpretation of the structures it reveals. Heidegger's renewal of ontology thus hinges on thinking about human existence in temporal terms and questioning traditional assumptions about the meaning of time. To these ends, Heidegger introduces the fundamental distinction between *time* as it is ordinarily understood and traditionally conceived, and *temporality* as the origin of that understanding. Heidegger complicates things significantly by distinguishing further between authentic and inauthentic temporality, temporality and intratemporality, now-time and world time, and the temporality (*Zeitlichkeit*) of Dasein and the Temporality (*Temporalität*) of being. Nevertheless, the primary distinction that Heidegger articulates in these various phenomena in *Being and Time*, division two, is the one he announces near the end of "The Concept of Time" when he enigmatically says, "Dasein is time, time is temporal" and immediately amends it with "Dasein is not time, but temporality" (CT 20/123). In the most basic terms, time is a "what," an entity, or a "how much," a quantity—or at least it has traditionally been conceived as such—whereas temporality is a "how," a way of being, or a "who," the being that says "I." Temporality, as the meaning of the being that we ourselves are, is the origin of time and the key to a "primordial interpretation" of the being of the subject, the soul, consciousness, the spirit, or the person—in short, human being.

In *The Metaphysical Foundations of Logic*, Heidegger credits Bergson with being the first to make a distinction between "original" and "derived" time (MFL 203/262). But the similarity between this distinction and Heidegger's own fundamental distinction between temporality and time is not obvious in *Being and Time*. It is more apparent in "The Concept of Time," where he describes time as "thoroughly uniform, homogeneous" (CT 4/110) and argues that "the original way of dealing with time is not a measuring" (CT 14/118). Claiming that irreversibility and homogeneity are what distinguish the concept of time from authentic temporality, he explains, "*Homogenization* is an assimilation of time to space, to Presence [*Präesenz*] pure and simple" (CT 18/121). In *Being and Time*, Heidegger speaks of "leveling down" and "covering over" instead of homogenization, but he makes the same point, already elaborated in great detail by Bergson: while the measurement of time renders every "now" the same, no two moments are alike in our experience. In *Being and Time*, Heidegger preserves this fundamental distinction between time and temporality, but he distances himself from Bergson by attempting to show that the problem with our understanding of time has everything to do with presence and little or nothing to do with space.

While Heidegger's distinction between time ("derived") and temporality ("originary") is analogous to Bergson's distinction between time and duration, Heidegger takes pains to avoid Bergsonian terminology. As Bergson describes them in *Time and Free Will*, duration is immediately experienced or *lived*, while time is represented in language and thought; duration is *real* and *concrete*, while time is artificial and abstract; duration is a *qualitative* multiplicity, while time is quantitative and discrete multiplicity; duration is *heterogeneous*, while time is a homogeneous medium; and duration is *pure* succession, while time is confused with juxtaposition in space. In *Being and Time*, Heidegger consistently avoids the term "duration," with the sole exceptions of his discussion of world time and his dismissal of Bergsonian duration. By placing duration at the level of world time, as an aspect of the "spannedness" of time, he attempts to show how it is less primordial than ecstatic temporality. Heidegger also refrains from describing temporality as "lived," "real," "concrete," or "pure," and he avoids all of Bergson's primary ways of characterizing duration: as a qualitative multiplicity, as pure heterogeneity, as the permeation of conscious states, as succession without mutual externality, as the flow of consciousness, and as a rhythmic organization of life. Heidegger radically simplifies Bergson's distinction by interpreting it as a difference between two kinds of succession, quantitative and qualitative, and by rejecting that distinction as a traditional one that blocks access to originary temporality. Nevertheless, Heidegger continues to maintain, as Bergson does, that both philosophy and common sense offer a distorted view of time, and he follows Bergson in seeking to uncover primordial time.

It is important, though, not to overstate the similarity between Bergson's challenge to the traditional concept of time and Heidegger's. Apart from any personal reasons he might have had for not giving Bergson his due, Heidegger distances himself from Bergson on the basis of significant philosophical differences. He regards Bergson as one of the leading figures in the philosophy of life, a movement he followed with great interest and criticized repeatedly in his early lectures and essays. In his efforts to promote phenomenology and develop a hermeneutics of "factical life," he is concerned about any philosophical treatment of life—including Nietzsche's, Dilthey's, and Jaspers's—that fails to overturn traditional presuppositions about the kind of being designated by "life." In this respect, Bergson's thought is also in the company of Scheler's and Husserl's, both of whom attempt to make a radical distinction between personhood and thingliness but remain within the limits of "anthropological" thinking. For Heidegger, these thinkers all continue to work within the ontic sciences of "life," "man," and "consciousness," rather than making the leap into ontology.

In addition, Heidegger thinks that while Bergson realizes that the traditional concept of time is problematic, he misdiagnoses the problem because he does not raise the question of being. Bergson's philosophy is ontologically inadequate, in his view, because it neither recognizes nor challenges the traditional interpretation of being as presence. Bergson argues that we have a distorted view of ourselves because we habitually represent time in terms of space, and that time as a homogeneous medium is an illegitimate concept. For this reason, Heidegger interprets Bergson as attempting to overcome the traditional concept of time by rejecting it as illusory, or as a confusion of duration with space. However, the problem, for Heidegger, is not just that we tend to think of time in terms of space, but that we interpret time as something objectively present. Heidegger dismisses Bergson's distinction between duration and time as a reversal of Aristotelianism that is insufficiently radical because it does not challenge the privilege of presence. This allows him to make a similar distinction without inviting comparisons that could overshadow his own insight that most of our perplexity about time is based on the assumption that being means being present.

It is easy to see how Heidegger holds Bergson up as a cautionary example of an errant way of thinking through time. For Bergsonian duration is, in Heidegger's view, still a form of temporality that privileges the present, a form of "now-time." Bergson's path is a dead end, for Heidegger, because originary temporality cannot be discovered simply by purifying the duration of conscious experience of any contamination by space. However, we can also see how Heidegger takes Bergson as a model for challenging traditional assumptions about time by distinguishing the way we usually represent it from a more primordial phenomenon of time. While Heidegger denies that duration is primordial time, he continues to follow Bergson in disentangling temporality from spatiality and in thinking about selfhood in temporal terms. In fact, Heidegger challenges the privilege of presence by interpreting Dasein as temporality and thoroughly distinguishing its being from anything spatial. Also, even though ecstatic temporality, as Heidegger characterizes it, is not a succession of conscious states, it is a kind of movement that is dimly reflected in the traditional concept of time. Finally, although he insists that the time we measure with clocks is a genuine phenomenon, he sees it as derived from inauthentic temporality. For better or worse, the very distinction between a concept of time derived from our needs and interests, on the one hand, and a more original experience of time, on the other, is deeply Bergsonian. It should therefore come as no surprise that even though Heidegger appears to have abandoned his plan for a decisive confrontation with Bergson, *Being and Time* is not the end of the story.

4

Reversing Bergsonism

Time and Temporality in *The Basic Problems of Phenomenology*

1. Thinking More Radically about Time

Toward the end of the footnote in *Being and Time* in which Heidegger dismisses Bergson, he postpones a more detailed critique with the comment, "This is not the place for a critical discussion of Bergson's concept of time and other present-day interpretations of time" (BT 410n/433n), adding a promissory note: "We shall come back to this in the first and third divisions of Part Two" (BT 433n).[1] Although no decisive confrontation with Bergson occurs in Heidegger's subsequent works, there are additional references to Bergson and his philosophy of time in his summer courses of 1927 (*The Basic Problems of Phenomenology*) and 1928 (*The Metaphysical Foundations of Logic*). In both works, as in his 1925–6 summer and winter courses (*History of the Concept of Time* and *Logic: The Question of Truth*), Heidegger acknowledges Bergson's importance in a way that he never does in *Being and Time*. In *Basic Problems*, he even fills in some of the details of the critique of Bergson sketched toward the end of *Being and Time*. During his lengthy discussion of Aristotle's account of time, Heidegger continues his engagement with Bergson and gives it a new twist: he argues not only that Bergson's understanding of time remains fundamentally Aristotelian, but also that Bergson's *misunderstandings* of Aristotle underlie his thesis that "time is space."

Heidegger's interpretation of Aristotle's concept of time, sketched in *Being and Time* and fleshed out in *Basic Problems*, serves not just as a critique, but also as an indication of untapped resources in Aristotle's thought. In *Being and Time*, as we have just seen, Heidegger's brief discussion of Aristotle's concept of time is limited to showing how his definition of it in the *Physics* makes "world time," which is dominated by making-present, explicit as "now-time." In this way, Aristotle raises the ordinary understanding of time to conceptual comprehension as a succession of "nows." Thus, Aristotle's definition of time, which is decisive for all subsequent philosophical discussions of the issue, produces a leveled-down version of ecstatic temporality. Understanding being as substance (*ousia*) or presence (*parousia*), Aristotle treats time and its components, the nows, as objectively present things, thereby covering over temporality and even world time with a "natural" interpretation of it.

In *Basic Problems*, Heidegger returns to this critique of Aristotle, going into much greater detail about the kind of experience on which his account of time is based, namely the experience of motion, and the assumptions about being that lead Aristotle to focus on the "now" as the essence of time. Heidegger emphasizes that this interpretation of Aristotle is phenomenological, focusing on how exactly time shows itself in Aristotle's concept, and ontological, bringing to light the understanding of being that guides Aristotle's investigation. In addition, Heidegger interprets Aristotle according to his strategy for the destruction of traditional ontology, which means not only criticizing or exposing the limits of Aristotle's thought, but also revealing untapped resources or hidden possibilities for positive appropriation. As such, Heidegger develops two readings, one emphasizing how Aristotle's treatment of time covers up the primordial phenomenon, and another indicating ways in which he uncovers the ecstatic and horizonal structure of temporality.

According to Heidegger, one of the aims of his exposition and interpretation of Aristotle's account of time is to clear up some common misinterpretations. Interestingly, the only name that Heidegger associates with these misinterpretations is Bergson's. As in *Being and Time*, he presents Bergson as a cautionary example, but in *Basic Problems* he cautions not only against confusing Bergsonian duration with originary temporality, but also against misunderstanding Aristotle as Bergson does. He continues to offer Bergson as an example of the predominance of Aristotle's concept of time and to argue that Bergson's attempt to overcome that concept is based on a mere reversal of Aristotle's thinking. Yet he goes further by claiming that Bergson misunderstands Aristotle's account of time in ways that lead to the view that Aristotle reduces time to space. Thus, the misunderstandings that Heidegger discusses are connected with Bergson's concern about thinking spatially about time.

Bergson's thought plays an even more important role in *Basic Problems* than in *Being and Time*, and one that has yet to be acknowledged. Heidegger uses his discussion of Aristotle's concept of time as an opportunity to work out his critique of Bergson. At the same time, he attempts to reveal resources in Aristotle for thinking about temporality by showing that Aristotle does not reduce time to space. To indicate certain hidden possibilities in Aristotle's thinking, Heidegger points out how the ecstatic-horizonal constitution of temporality shows through Aristotle's concept of time in the *Physics*. On this interpretation, even though Aristotle understands time in terms of the "now," he catches *glimpses* of originary temporality. Heidegger shows this is by defending Aristotle against the Bergsonian concern about confusing time with space, arguing that Aristotle does not think about time in terms of space. Paradoxically, Heidegger thus portrays Aristotle as thinking more radically about time—having a clearer view of the structure of temporality—than Bergson. However, as I will argue, it is only by means of certain inversions and omissions in his reading of Aristotle that Heidegger manages to portray his thinking about time as closer to originary temporality than Bergson's.

I argued in Chapter 2 that Heidegger's dismissal of Bergson in *Being and Time* relies on an oversimplification of his thought and in Chapter 3 that Heidegger's attempts to distance himself from Bergson reveal their proximity, as Heidegger appropriates some of Bergson's key insights about time. With regard to *Basic Problems*, I aim to show that Heidegger continues to engage with Bergson in an indirect but significant way, again by subordinating his thinking about time to Aristotle's. First, I consider Heidegger's comments about Bergson and the traditional concept of time at the beginning of his discussion of temporality. Heidegger reiterates both his reasons for taking Bergson seriously, which appear in previous lecture courses but are omitted from *Being and Time*, and his reasons for dismissing Bergson, again with reference to Aristotle's concept of time. Then I examine Heidegger's remarks in the course of his interpretation of Aristotle's account of time about what Bergson misunderstands and how it leads to the misconception that Aristotle reduces time to space. I show how Heidegger reads Aristotle not only as formulating the traditional concept of time, which covers up originary temporality, but also as catching a glimpse of the ecstatic-horizonal structure of temporality. Finally, I criticize the selective reading of Aristotle that Heidegger uses to interpret him as having already anticipated Bergson's worry about confusing time with space. With this interpretation, Heidegger both confines Bergson's thought to the tradition inaugurated by Aristotle and uses Bergson's critique of that tradition as leverage to lift Aristotle above standard interpretations. The result is a strangely inverted history of the concept of time, in which Bergson appears as being thoroughly traditional and Aristotle as providing phenomenological insight into temporality that is more radical than those of Heidegger's own contemporaries.

2. Destruction of the Traditional Concept of Time

Heidegger introduces *The Basic Problems of Phenomenology* as "a new elaboration of division three of part one of *Being and Time*" (BPPb 1n/1n).[2] Insofar as Heidegger's interpretation of Aristotle had been planned for part two, division three of *Being and Time*, his examination of Aristotle's concept of time in *Basic Problems* also makes headway toward the third stage of his destruction of the history of ontology. In §19 of *Basic Problems*, Heidegger begins with a close reading of *Physics* IV, 10–14, where Aristotle defines time and responds to various problems with it, setting the agenda, in Heidegger's view, for the entire Western philosophical tradition's approach to time. Heidegger's reading expands on the one in *Being and Time*, §81, by explicating Aristotle's account of time in a more detailed and methodical way and by explaining in what ways it is definitive for subsequent philosophers. One of Heidegger's aims is to show more clearly how the traditional concept of time, and thus the ordinary understanding of time, arises from and obscures originary temporality. Yet Heidegger does not treat Aristotle's definition of time merely as something to be rejected or superseded. "No attempt to get behind the riddle of time can permit itself to dispense with coming to grips with Aristotle" (BPPb 232/329), he insists, for Aristotle is the thinker who first raises the ordinary understanding of time to the level of a concept. If that understanding of time indeed arises from what Heidegger now calls "ecstatic-horizonal temporality," then the task is to reveal how and to what extent the structure of temporality shows through the concept of time handed down through the history of philosophy.[3]

At the beginning of part two of *Basic Problems*, Heidegger provides a compendium of primary sources on the philosophy of time. The classical texts he mentions include Aristotle's *Physics* and *De Anima*, Plotinus's *Enneads*, and Simplicius's commentary on the *Physics*. From the medieval and modern periods, he names Aquinas and Suárez, passing over Augustine only because he had begun his discussion, as he also had in "The Concept of Time," with the famous passage from the *Confessions* in which Augustine puzzles over how time is at once familiar and inexplicable. Heidegger lists Leibniz, Kant, and Hegel as leading representatives of modern philosophical inquiries into time. Turning to contemporary thought, he writes: "From the most recent period we may cite Bergson's investigations of the time phenomenon. They are by far the most independent. He presented the essential results of his inquiries in *Essai sur les données immédiates de la conscience*" (BPPb 231/328).[4] However, before this praise for Bergson's work has a chance to sink in, Heidegger continues:

> As early as his first treatise, Bergson makes the attempt to overcome the Aristotelian concept of time and to show its one-sidedness. He tries to get beyond the common [*vulgäre*] concept of time by distinguishing *durée*, duration, in contrast with time as ordinarily understood, which he calls *temps*. . . . Bergson's theory of duration itself grew out of a direct critique of the Aristotelian concept of time. The interpretation he gives of time in the common sense rests on a misunderstanding of Aristotle's way of understanding time. Accordingly, his counter-concept to common time, namely duration, is also in this sense untenable. He does not succeed by means of this concept in working his way through to the true [*eigentlich*] phenomenon of time. Nevertheless, Bergson's investigations are valuable because they manifest a philosophical effort to surpass the traditional concept of time. (BPPb 232/328–9)[5]

By now, most of this sounds like a refrain from Heidegger's 1925–6 lecture courses and *Being and Time*. Heidegger again characterizes Bergsonism as a failed attempt to go beyond Aristotle and get to the root of time. The *effort* to go beyond the traditional concept of time constitutes the value of Bergson's investigations, but insofar as he conceives duration in opposition to Aristotle's concept of time, Heidegger argues, his insight is limited by the ontological orientation of ancient Greek thought. Heidegger thus signals that his own interpretation of Aristotle will avoid the pitfall awaiting those, like Bergson, who attempt to bring the phenomenon of time into view without coming to terms with traditional ontology.

The pitfall Heidegger aims to avoid by returning to Aristotle's *Physics* is that of professing to think about time in a radically new way while actually being under the sway of assumptions about being that have dominated philosophy since the ancient Greeks. Early in *Basic Problems*, in an introduction to the phenomenological method as he employs it, Heidegger describes that method as having three components: reduction, construction, and destruction. Heidegger uses the reduction, Husserl's method for arriving at pure consciousness, to turn attention "back from the apprehension of a being, whatever may be the character of that apprehension, to the understanding of the being of this being" (BPPb 21/29). That understanding is already given—Dasein understands being and projects itself understandingly—so construction is the projection of that being in an ontologically adequate way. For Heidegger, this means rethinking being altogether. But here he emphasizes the need for a *historical* destruction because "all philosophical discussion, even the most radical attempt to begin all over again, is pervaded by traditional concepts and thus by traditional horizons and traditional angles of approach" (BPPb 22–23/31).[6] Destruction is, he says, "a critical process in which the traditional concepts, which at first must be necessarily employed, are deconstructed down to the sources from which they were drawn" (BPPb 22–23/31). This means going back through the "history of ontology," as Heidegger had proposed in *Being and Time*, and revealing the limits of the philosophical tradition, that is, the ways it is limited by the presuppositions of the thinkers who shaped it. Through this process, Heidegger intends to show the extent to which basic concepts are genuinely derived from "primordial experiences" and the extent to which they are determined by traditional ontology.

Heidegger stresses that his attitude toward traditional concepts and his pursuit of their deconstruction are not purely critical. Phenomenological destruction is by no means a rejection of traditional concepts, but, "quite the reverse, it signifies precisely a positive appropriation of tradition" (BPPb 23/31).[7] Philosophy is never a matter of turning our backs on tradition and making a fresh start; this is neither desirable nor even possible, for Heidegger. The resources for thinking differently are there in the tradition, hidden by prejudices, presuppositions, and concepts accepted as self-evident. What is necessary is to go back to the experiential origins of our concepts, bring them to light phenomenologically, mark the limits of their formation within the prevailing understanding of being, and "stake out the positive possibilities in that tradition" (BT 22/22). Thus, while destruction is critical insofar as it is concerned with the limits of the philosophical tradition, this criticism is always at the service of uncovering and appropriating untapped resources. Phenomenological destruction is concerned above all with revealing hidden possibilities for new ways of thinking.

With respect to time, the destruction of ancient Greek ontology begins with Aristotle because, in Heidegger's estimation, "he expressed in clear conceptual form, for the first time and for a long time after, the common understanding of time, so that his view of time corresponds to the natural concept of time" (BPPb 232/329). Instead of trying to overcome the traditional concept of time, as Bergson attempts and fails to do, Heidegger's aim is to trace it, as first formulated by Aristotle, back to the forgotten experiences from which it arose. The question is whether and to what extent Aristotle's concept of time "reaches the phenomenon in its original constitution" (BPPb 230/326). Heidegger's interpretation of Aristotle takes place in two stages: first, an outline of Aristotle's account of time in *Physics* IV, chapters 10 to 14, and second, an "interpretive exposition" of his concept of time, focusing on "the phenomenon as Aristotle sees it" (BPPb 237/336) and the ontological meaning of his definition. His reading of Aristotle here proceeds according to his previous characterization of phenomenological destruction. On the one hand, he presents Aristotle as "the last of the great philosophers who had eyes to see" (BPPb 232/329) the phenomenon of time. On the other hand, he claims that Aristotle's approach raises great difficulties thanks to its ontological presuppositions. If Heidegger seems to vacillate, praising Aristotle and defending him against critics at one moment, then attacking him the next, it is because he wants to illuminate not only the limits of Aristotle's thinking, but also the dimensions of his thought that have remained hidden.[8] Heidegger thus provides one reading that shows how Aristotle's concept of time as a sequence of nows covers up world time and levels down temporality, as he had argued in *Being and Time*, and another one that shows how Aristotle's definition of time and his analysis of the now offer glimpses of originary temporality.[9] To understand the significance of Heidegger's remarks about Bergson in *Basic Problems*, it is necessary to recognize this double reading, in which Heidegger both marks off the limits of Aristotle's thought and explores other possibilities for radically rethinking time.

3. A Glimpse of Temporality

The limits of Aristotle's concept of time can already be seen in the questions with which he begins and ends his account of time in the *Physics*. He begins with two questions: "does time belong among beings or non-beings?" and "what is the nature, the essence, of time?"[10] With the first question, Aristotle asks about "*time's mode of being*" (BPPb 233/330). This question is not easily answered, for the past *is* no longer, the future *is* not yet, and the present—the "now"—lacks the self-sameness and unity of something that exists. The *aporia* at which Aristotle leaves the question after a very brief discussion is this: How could something composed of nonexistent parts exist? Aristotle's second question, which is about time's "*essential nature*," receives more of his attention. It is this question that leads Aristotle to define time as the "*arithmos kinēseos kata to proteron kai husteron*,"[11] which Heidegger translates somewhat idiosyncratically as "something counted in connection with motion as encountered in the horizon of earlier and later" (BPPb 235/333). Another key question arises toward the end of Aristotle's account, namely whether time is subjective. On the one hand, if time is indeed "something counted," and counting is an operation that takes place "in the soul," then it may be that time itself exists only in the soul or mind.[12] On the other hand, if time is something connected with motion, then perhaps it is as objective as motion is. Both this question and Aristotle's initial question about time's mode of being remain unresolved, although he presumes that time exists while investigating its essential nature. For Heidegger, these *aporia* illustrate not just the difficulty of understanding time, but also the limitations of the ontological framework with which Aristotle operates.

These questions perplex Aristotle, as well as his philosophical successors, thanks to his understanding of being as *ousia* (substance) or *parousia* (presence), according to Heidegger. While Aristotle identifies first philosophy with ontology, raising the question of the meaning of being that unifies all the ways in which something can be said to be, Heidegger argues that he is not fully aware of the presuppositions about time involved in his understanding of being. Specifically, he does not seem to recognize the primacy of the present implied in the interpretation of being as presence. "Aristotle says that time is *kinēseos ti*, something *connected with* motion" (BPPb 272/385), but there are difficulties surrounding the "is." His ontological orientation, as Heidegger has already emphasized in *Being and Time*, privileges the "extant" or "objectively present" (*vorhanden*).[13] Heidegger writes: "If the common understanding of time is aware of being only in the sense of objectively present being, then time, being publicly accessible *along with* motion, must necessarily be something objectively present" (BPPb 272/385, translation modified). If being means presence, as Heidegger explains in the introduction to *Being and Time*, then whatever *is* must be present in some manner.[14] This interpretation of being—and of access to it through speech (*legein*) and intuition (*noein*), privileged ways of making-present—equates it with being present in time. But how could time as a whole be present in time? The difficulties faced by Aristotle and passed down through the history of philosophy stem from the paradox that being is understood in terms of a particular mode of time, the present, and time is understood on the basis of that interpretation of being.[15]

The questions of whether time exists and whether it is subjective turn out to rest on the assumption that if time exists, it must be an objectively present thing. This directs Aristotle's attention to the "now," for "[t]he time that is known as the now and as a manifold and succession of nows is an *objectively present sequence*" (BPPb 272/385, translation modified). It follows from the interpretation of being as presence that the now must be the only aspect of time that truly exists, because the past and the future are absent by definition (being "no longer" and "not yet" being). For this reason, as Heidegger had argued in *Being and Time*, Aristotle's concept of time deserves to be called "now-time." The attempt to understand time in terms of presence is the source of great confusion, for "Aristotle's *aporia* with reference to the being of time—which is still the principal difficulty today—derives from the concept of being as equal to being objectively present" (BPPb 272/385, translation modified). While time as a whole can be represented as a sequence or succession of nows, only one now can be present in actuality. Aristotle is not willing to conclude that time is subjective, but some of his most influential successors are. Augustine, for instance, saves the past and future from nonexistence by regarding their existence as mental: the past exists as present in memory, and the future exists as present in expectation. Similarly, Kant defines time as the form of inner sense, an *a priori* condition of experience, which organizes all appearances but cannot be attributed to things in themselves. According to Heidegger, to the extent that the entire philosophical tradition follows Aristotle in treating time as "objectively present, just as space is," the limits of his concept of time are also those of "every subsequent account of time, including Bergson's" (BPPb 272/385).

That said, Heidegger also builds a case—and this constitutes his second reading—that Aristotle's elucidation of "*the common prescientific understanding of time*" nevertheless offers a glimpse of "an original time, *temporality*" (BPPb 257/362). On this reading, the structure of originary temporality appears in filigree, so to speak, in Aristotle's treatment of time. Even though "Aristotle characterizes time primarily as a *sequence of nows*" (BPPb 256/362), his definition provides access through "now-time" or the "time of nature" to ecstatic and horizonal temporality. With his paraphrase of Aristotle's definition of time as "something counted in connection with motion as encountered in the horizon of earlier and later," for example, Heidegger draws attention to the *horizonal* structure of temporality. He argues that, contrary to the common objection that Aristotle's use of temporal terms in his definition of time renders it circular or tautological, the "earlier and later" is really evidence of "the inner coherence of the Aristotelian time phenomenon, that is, of time as commonly understood, with the original time we are calling temporality" (BPPb 241/341). For Heidegger, the structure of originary temporality is not merely covered up by Aristotle's concept of time, but it shows through it as well.

Not only does his definition of time allow the horizonal structure of temporality to show through, on this interpretation, but Aristotle's characterization of the now reveals the *ecstatic* character of temporality as well, because the now "has dimension within itself; it stretches out toward a not-yet and a no-longer" (BPPb 248/351). Aristotle's account of what Heidegger calls the "peculiar double visage [*eigentümliches Doppelelgesicht*]" (BPPb 247/349–50) of the now reflects both the unity of time and its division into the not-yet and the no-longer. Aristotle recognizes that the phenomenon of the now is a strange one, because the past and present are continuously bound together by it, but by that very fact it seems to be sundered or at least pulled in two directions. Aristotle sees that the now has to "stretch out" for motion (and rest) to be accessible as such and to be measured in terms of time. Insofar as the now stretches out, it has "dimension" and therefore cannot be identified with a spatial point. In Heidegger's view, the now thus exhibits the ecstatic character of temporality, the "*primordial 'outside of itself' in and for itself*" (BT 314/329). That character may be leveled down in the concept of time as a sequence of nows, but it appears in the distinctive structure of the now that Aristotle describes as not just a limit between past and future, but also a transition that makes the experience of motion possible.[16]

In line with Heidegger's characterization of phenomenological destruction, his interpretation of Aristotle thus comprises a double reading that draws attention to both the limits of Aristotle's thinking, or what prevents him from reaching originary temporality, and the resources available for developing new interpretations and revealing hidden structures. With this reading, Heidegger maintains his position in *Being and Time* that Aristotle's concept of time presupposes "the 'natural' way of understanding Being" (BT 400/421), but goes on to argue that contemporary readers of Aristotle, with Bergson as a prime example, overlook the extent to which Aristotle at least catches a glimpse of originary temporality through now-time. In this respect, it is incorrect to regard Aristotle's concept of time as having completely covered up or distorted the primordial phenomenon. What Heidegger argues is that while Aristotle's view of time is indeed limited by his understanding of being as presence, it is not open to the Bergsonian objection that Aristotle reduces time to space. That objection, he claims, suffers from several misunderstandings of Aristotle's definition of time and his explication of the now, which contribute to Bergson's failure to provide an undistorted view of temporality.

4. Bergson's Misunderstandings

Heidegger's remarks about Bergson in the context of his interpretation of Aristotle in *Basic Problems* signal an interesting shift in Heidegger's critique. Now the problem is not just that Bergson's interpretation of time is essentially determined by his understanding of Aristotle, but also that it is determined by several fundamental *mis*understandings. As we have seen, in *Being and Time* Heidegger focuses on Bergson's concept of duration as a "qualitative succession" and his attack on the concept of time as a spatial representation of duration. According to Heidegger's critique, Bergson characterizes duration in direct opposition to Aristotle's concept of time by arguing that lived time is a qualitative succession rather than a quantitative one. This is merely a reversal of Aristotle's definition of time as something counted, a kind of number, connected with motion. Moreover, Bergson claims that "time is space," meaning that the way we represent time as a homogeneous medium confuses it with space. For Heidegger, Bergson's analysis is an indictment not just of the way we ordinarily understand time, but also of the concept of time formulated by Aristotle. However, while Heidegger's discussion of Bergson in *Being and Time* is limited to making the connection between Bergson's understanding of time and Aristotle's, in *Basic Problems* he goes further by challenging the interpretation of Aristotle that appears to motivate Bergson's reversal. Heidegger does so by attempting to show that Aristotle can only be read as thinking spatially about time, or failing to adequately distinguish between time and space, on the basis of several misunderstandings. With each misunderstanding that Heidegger attributes to Bergson, he attempts to show how Aristotle distinguishes time from space.[17]

Bergson's first alleged misunderstanding, which is at the root of the misconception that Aristotle reduces time to space, involves a connection that Aristotle makes between time, motion, and extension. He defines time on the basis of the experience of motion, and this misunderstanding begins with an overly narrow interpretation of motion. For Aristotle, time is "not in one definite place, and it is not in the moving thing itself"; nevertheless, "time also does not exist without motion" (BPPb 235/333). While demonstrating the link between time and motion, Aristotle disagrees with those who identify time with a particular kind of motion, namely that of the heavens. The question of time's connection to motion leads Aristotle to define time as something counted *in connection with motion* as encountered in the horizon of earlier and later. With regard to the continuity of time, Aristotle also claims that "time follows motion" and "motion follows extension" (BPPb 243/344). As such, not only is time continuous like space, but it is also a magnitude, something measurable. According to Heidegger, the view that Aristotle reduces time to space is based first of all on a misunderstanding of Aristotle's concept of motion and its connection with extension.

For Heidegger, Aristotle's concept of motion (*kinesis*) in general must not be understood too narrowly. When discussing time, Aristotle often uses local motion (*phora*) in his examples, but motion in the broadest sense includes such things as qualitative change (*alloiosis*). When something changes color, for example, this is an instance of *kinesis*, that is, "an advance *ek tinos eis ti, away from something toward something*" (BPPb 242/343). Heidegger elaborates that "this 'away from something toward something' does not have the sense of transition from one place to another" and "need not necessarily be taken spatially" (BPPb 242/343).[18] Instead, *kinesis* means change in general (*metabole*) and refers to any phenomenon with the basic structure of "away from something toward something" common to change of place, change of quality, and even change of mind.[19] Heidegger refers to this structure of the "away-from-something-toward-something" as *dimension*, a term that "expresses a general notion of *stretch*" (BPPb 242/343).[20] Dimension, as Heidegger presents it, is an ontological structure of which spatial extension is a particular mode, and in his view it is with regard to this structure that Aristotle's propositions "time follows motion" and "motion follows extension" must be interpreted:

> In the case of the determination of *ek tinos eis ti* we should rid ourselves completely of the spatial idea, something that Aristotle did, too. A completely formal sense of stretching out is intended in "from something to something." It is important to see this, because it was with reference to this determination that the Aristotelian concept of time was misunderstood in the modern period, especially by Bergson: from the outset he took this dimensional character of time in the sense of spatial extension in its reference to motion. (BPPb 242/343–4)[21]

Thus, if Bergson accuses Aristotle of reducing time to space, it is because he misinterprets Aristotle as restricting motion to local motion. Not only that, but he also misinterprets Aristotle with respect to extension or magnitude (*megethos*). For, just as *kinesis* includes more than local motion, *megethos* "has a broader sense than specifically spatial dimension" (BPPb 242/343). For Heidegger, the notion that Aristotle reduces time to space or confuses duration with extension is based on the misinterpretation of both motion and extension.

Heidegger contends that Bergson also misunderstands the nature of the connection that Aristotle establishes between extension, motion, and time. Aristotle says "*akolouthei to megethei he kinēsis*, motion follows (comes in the wake of) dimension (extension)" (BPPb 243/344).[22] By claiming that time is a number connected with motion, and motion follows extension, Aristotle could be read as reducing time to space. But in what sense do these phenomena "follow" one another? In Heidegger's view, Aristotle does not seek to establish a causal or logical connection between time, motion, and extension, but an ontological one: insofar as the formal structure of dimension or stretch implies continuity (*suneches*), "follows" refers to "the *foundational a priori connection of motion* with *continuity* and *extendedness*" (BPPb 243/344). That is, extension (qua "stretch" in general) is implicit in motion (qua change in general), which is implicit in time. If Aristotle assigned an order of ontic priority to these terms, then time would indeed be reduced to space. But if "follows" signifies an *a priori*, ontological connection between the phenomena, then it is wrong to charge Aristotle with reducing time to space. This is crucial because, Heidegger says, "[u]nless the ontological sense of *akolouthein* has been comprehended, the Aristotelian definition of time remains unintelligible. Or else defective interpretations occur, for example that of Bergson, who says that time as Aristotle understands it is space" (BPPb 244/345).[23] Thus, compounding the problem of interpreting motion and extension too narrowly is the problem of interpreting their connection ontically.

The second misunderstanding Heidegger attributes to Bergson deals with the "before and after" in Aristotle's definition of time. Contrary to Bergson's allegations about the way time is represented, Heidegger argues, time is no more reducible to space for Aristotle than motion is reducible to a series of places traversed by a moving thing. As Aristotle observes, any experience of motion requires that the moving thing appear in transition from one location or state to another. Thus, not even local motion is reducible to a multiplicity of points: "Only when we see the moving thing in its changing over from there to here do we experience motion, transition. . . . We must see the presented contexture of places, the point manifold, in the horizon of an 'away from there—toward here'" (BPPb 245/346–7). On this interpretation, Aristotle does not think of motion as constituted by a "point manifold" of indifferent, homogeneous places. The experience of motion is not an "adding up" or "reckoning" of points, but the retention of that-away-from-which-something moves (or changes) and expectation of that-toward-which it moves. In other words, motion is experienced as such always in the *horizon* of an earlier and a later.

However, the *proteron* and *husteron* in Aristotle's definition of time raises a well-known dilemma for interpreters. On the one hand, these terms can be translated as "earlier" and "later," but because they denote temporal characteristics, doing so would seem to render Aristotle's definition of time "a trivial tautology: time is the earlier and later, thus *time is time*" (BPPb 241/341). On the other hand, they can be translated as "before" and "after" to avoid the tautology if these terms are taken to refer simply to relative position in a series (as, for instance, the number four is before five and after three). That is, they can be taken in a spatial sense rather than a temporal one, removing the apparent circularity in Aristotle's account. Many interpreters take this route, and Aristotle even says, "The distinction between before and after holds primarily, then, in place."[24] The problem is that interpreting *proteron* and *husteron* spatially in order to avoid a circular or tautological definition of time thereby reduces time to space. Heidegger suggests a different route, reinterpreting the apparent tautology "time is time," which results from taking *proteron* and *husteron* in a temporal sense: "Perhaps the second term 'time' means something different and more original than what Aristotle means in the definition itself" (BPPb 241/341). More precisely, Heidegger suggests, the "earlier" and "later" refer not to time in the ordinary sense, but to *originary temporality*. For Heidegger, not only Aristotle's definition of time but also his whole investigation should be understood as *deriving time from ecstatic-horizonal temporality rather than space*. What Bergson and other commentators have failed to see, in his view, is that a more primordial phenomenon of time shows through Aristotle's account of time as it appears in the experience of motion.

A third misunderstanding, for Heidegger, is to think of the Aristotelian "now" as nothing more than a spatial point. Whereas Bergson criticizes the idea that time consists of a succession of discrete moments and our habit of thinking of time as a series of points, Heidegger reads Aristotle as already resisting the tendency to represent time in this manner. To see that "Aristotle does not reduce time to space nor does he define it merely with the aid of space," we need to pay close attention to Aristotle's phenomenology and "recognize what is already experienced in and with the experience of motion and how time becomes visible in what is thus experienced" (BPPb 244/345). With the aim of clarifying the phenomenal content of Aristotle's definition of time, Heidegger discusses four constitutive moments of the experience of motion: 1) motion is never given in itself without *something* moving; 2) the moving thing undergoes *transition*, so motion always occurs along a continuum; 3) the moving thing undergoes transition *from there to here*, so motion is structured according to a horizon (indicated by the *proteron* and *husteron*); and 4) the moving thing that undergoes transition from there to here is experienced in *the now*.[25] The now is implicit in the experience of motion as Aristotle presents it, and the measurement of time is a matter of "counting nows." The privilege that Aristotle bestows on the phenomenon of the now explains why Heidegger calls his concept of time "now-time." But is the Aristotelian now nothing more than a point through which things pass?

According to Heidegger, when Aristotle describes the now as stretching toward both the past and the future, he distinguishes it fundamentally from a spatial point. For Aristotle, the now exhibits a peculiar twofold structure, which Heidegger calls its "double visage." Appearing as it always does in the horizon of the earlier and the later, the now functions both to bind time and to sunder it. As Heidegger explains, "time's specific continuity is rooted in the now. But conjointly, with respect to the now, time is divided, articulated into the no-longer-now, the earlier, and the not-yet-now, the later" (BPPb 247/350).[26] The peculiarity of the now is thus that it simultaneously constitutes both a *transition* from earlier to later and a *boundary* between them.[27] Put differently, the now is characterized by both alterity and identity, which Aristotle expresses in a paradox: "The ever different nows are, *as different*, nevertheless always exactly *the same*, namely, now" (BPPb 247/350). So "now" refers ambiguously to both a fleeting instant, distinct from every other instant, and a constant index of the past as no longer and the future as not yet present. Janus-faced, linking the past and the future while marking the difference between them, and maintaining its identity while always changing, the Aristotelian now is therefore fundamentally distinct from anything objectively present.

Aristotle's recognition of this distinctive structure of the now, which he says is always the same *and* always different, prevents him from confusing moments of time with spatial points, in Heidegger's view. The problem only arises when we forget about the twofold character of the now and represent it as objectively present together with other nows. To understand Aristotle's interpretation of time, he claims, "it is not enough that we correlate the nows in juxtaposition to a point-manifold, so as to think of them as being at a standstill in a line. This talk of time as a sequence of nows should not be misunderstood and transferred to the *spatial*, thus leading us to speak of time as a line, a series of points" (BPPb 248/350). Contrary to a common misinterpretation, Aristotle avoids this error by revealing the "stretching out within itself" exhibited by the now—its ecstatic character—and by stressing its function as a transition. According to Heidegger, Aristotle maintains that the now functions as a limit "only accidentally" and that its "essential nature" is openness toward both the future and the past (BPPb 249/352). This openness makes it possible to experience motion and to measure it by counting nows, for in every now in which we count, we must both retain former nows and await upcoming ones. The now, for Aristotle, "is already always beyond the point . . . it looks backward and forward . . . it is beginning and end" (BPPb 250/354). To interpret Aristotle as reducing time to space is thus, in Heidegger's view, to overlook this distinctive structure of the now as the condition for the possibility of not only experiencing motion, but also measuring it.

A final thing that Bergson misunderstands, according to Heidegger, is the meaning of Aristotle's association of time with number. Aristotle defines time as a kind of number, or something counted, but in what respect, exactly, is time similar to number? Heidegger notes that just as number is independent of things that are counted, time is independent of the "what" (or "*intrinsic content*") and the "how" (or "*mode of being*") of things that change (BPPb 250/353). Because time is independent of what it measures, in a sense it stands outside of things; conversely, being "in time" simply means being measurable in temporal terms: "According to Aristotle, 'things are in time' means nothing but that they are measured by time on the basis of their transitional character" (BPPb 251/355). Whereas Bergson attacks the metaphor of time as a container, regarding it as a spatial representation, Heidegger defends Aristotle's notion of time as containing—or, in his translation, "embracing" (*umgriffen*)—things in motion and at rest (BPPb 252/356). In Heidegger's interpretation, this is a formal indication of the structure of time as the horizon in which motion is experienced. It does not imply a spatial relationship between time and things in time, but a particular mode of being discussed in *Being and Time*: intratemporality.[28]

Not only does Aristotle's association of time with number *not* imply a reduction to space, for Heidegger, but it also sheds light on the difference between moments of time and objectively present things, namely that they are "in time" in a completely different sense. Whereas the nows *constitute* time, according to Aristotle, things in motion or at rest *belong to* time "in the way in which what is counted belongs to number" (BPPb 252/356). Whereas Bergson objects to the metaphor of containment as inappropriate for understanding time, Aristotle regards time as embracing everything measurable by it. Heidegger does not mention Bergson explicitly during this discussion of intratemporality in *Basic Problems*, but a comment from *Being and Time* shows the extent to which his analysis of intratemporality is a response to Bergson. There, Heidegger writes, "the time 'in which' objectively present things come into being and pass away is a genuine phenomenon of time; it is not an externalization of a 'qualitative time' into space, as Bergson's interpretation of time—which is ontologically completely indeterminate and insufficient—would have it" (BT 318/333). Heidegger now suggests that Aristotle's association of time with number clarifies the phenomenon of intratemporality. Anything in motion or at rest is said to be "in time," whereas things that never change (such as geometrical figures and their properties) are considered "timeless," that is, "neither embraced nor embraceable by time" (BPPb 253/357). Aristotle's account of time thus provides a key to understanding the traditional ontological function of time as a "criterion for naïvely distinguishing the different regions of beings," those of the temporal and the timeless (BT 18/18). The association of time with number seems to be misunderstood by Bergson, who mistakes Aristotle's insight into the character of time as an "embracing horizon" for a confusion of time with space. As with what Heidegger calls "dimension," embracing is a formal structure that is ontologically prior to the distinction between time and space.

In summary, Heidegger accuses Bergson of misunderstanding Aristotle's account of time in four ways. First, he misunderstands what Aristotle means by "motion follows extension," incorrectly assuming that "motion" and "extension" refer narrowly to local motion and spatial extension, rather than broadly to change in general and stretching out. He also misunderstands the link that Aristotle establishes between the phenomena of extension, motion, and time, which Heidegger interprets as an ontological connection. Second, Bergson thinks that "before and after" must refer primarily to place in order to prevent Aristotle's definition of time from being tautological, whereas "earlier and later" would better indicate time's horizonal character. Third, Bergson overlooks the difference that Aristotle sees between the now and a spatial point, with the now having a distinctive ecstatic structure. And fourth, insisting that it is no spatial container, Bergson denies a fundamental aspect of time that Aristotle recognizes, namely the embracing character that it shares with number. In Heidegger's view, these misunderstandings are responsible for the deficiency of Bergson's interpretation of Aristotle. Furthermore, if Bergsonian duration is indeed an outgrowth and a reversal of Aristotle's concept of time, then these misunderstandings provide insight into why Bergson does not get to the root of time.

5. De-spatializing Aristotle's Thinking about Time

Throughout *The Basic Problems of Phenomenology*, Heidegger takes it for granted that Bergson accuses Aristotle of reducing time to space. As I have argued, however, Heidegger's evidence that Bergson's concept of duration arises from his interpretation of Aristotle's concept of time is not particularly convincing. That evidence includes the appearance of Bergson's *Time and Free Will* around the same time as his essay "The Concept of Place in Aristotle," the preliminary analysis of number in the second chapter of *Time and Free Will* leading to his discussion of duration, and the use of the categories of quality and quantity to distinguish duration from time. The two theses at the center of Heidegger's critique—that "time is space" and duration is "qualitative succession"—oversimplify and even distort Bergson's thought. For Bergson actually develops his concept of duration in opposition to a habit of representation that he believes underlies most if not all philosophical and scientific conceptions of time. Moreover, he specifically criticizes Kant's account of time in the Transcendental Aesthetic while making no reference to Aristotle's account in the *Physics*. Consequently, when Heidegger argues that Bergson is incorrect to accuse Aristotle of reducing time to space, we should ask whether any such accusation has been made. That said, if we put aside these reservations about the very association of Bergson with Aristotle and examine what Heidegger says Bergson misunderstands, we can see a number of other difficulties with his arguments.

Several of the misunderstandings that Heidegger attributes to Bergson in *Basic Problems* involve the interpretation of some of Aristotle's basic concepts. Heidegger claims that, because of narrow interpretations of *kinesis* and *megethos*, Bergson overlooks the formal structure of dimension, something more fundamental than extension in space, and the ontological character of the link between time, motion, and extension. Heidegger has no trouble showing that Aristotle employs *kinesis* broadly, referring not just to change of place, but also to "transition from something to something," thanks to Aristotle's application of the term to qualitative change and even "movement of the mind." Heidegger applies the same reasoning to the term *megethos*, but in that case his only textual support is the ambiguity of the phrase "from something to something." He offers "dimension" as an alternative to "magnitude" or "extension" to indicate the "completely formal sense of stretching out" that Bergson and other commentators have interpreted spatially. Unfortunately, while it is clear that Aristotle employs *kinesis* in a broader sense than local motion, it is unclear whether this non- or pre-spatial sense of dimension is really a feature of Aristotle's account. One reason to suspect that it is not is that Aristotle says nothing to discourage the association of *megethos* with spatial extension. Heidegger's attempt to break this association is inventive, but his evidence is lacking.

Heidegger is also on shaky ground when he proposes that Aristotle employs *akolouthein* in an ontological sense with respect to the connection between time, motion, continuity, and extension. The claim that this term signifies an ontological connection is central to Heidegger's reading of Aristotle on time. But what justifies this interpretation, not to mention the bold assertion that Aristotle's concept of time is bound to be misunderstood without it? Perhaps the basis for an ontological interpretation is Aristotle's remark in *Metaphysics*, to which Heidegger alludes early in *Basic Problems*, that the fundamental question of philosophy is "What is being?" (BPPb 15/19). Heidegger uses this remark to confirm his own orientation, which becomes explicit when he defines philosophy as "the science of being" and phenomenological reduction as "the leading of our vision from beings back to being" (BPPb 21/29). From the standpoint of ontology, Heidegger argues that Aristotle reverses the order of phenomena with the claim "motion follows extension" because he is led back from the experience of motion to its *a priori* condition, the dimension or stretch of ecstatic temporality. In fact, it is Heidegger who reverses Aristotle's order of priority with this interpretation by regarding motion, continuity, and extension as "thought along with time" rather than regarding time as being derived from motion and motion from extension.[29] Because the traditional interpretation of the connection between these phenomena invites the criticism that time is thereby reduced to space, Heidegger defends Aristotle by interpreting his account of time as a clarification of ontological structures implicit in the experience of time. We should ask, however, whether Heidegger is drawing these structures out of Aristotle's account of time or reading them into it.

Heidegger's solution to another problem, the problem of the circularity created by the *proteron* and *husteron* in Aristotle's definition of time, is both ingenious and audacious. Not only does the translation "before and after" avoid the problem of using temporal terms to define time, but Aristotle also explicitly says that the distinction between *proteron* and *husteron* refers to place.[30] Heidegger concedes that the standard translation avoids making Aristotle's definition into the tautology "time is time," and he admits that "earlier and later" is a translation that "already introduces a large element of interpretation" (BPPb 246/349). Regardless, Heidegger prefers "earlier and later," provided that the phrase is understood transcendentally, as indicating the horizon in which nows are counted. This means that "earlier and later" refers to neither space nor time, but the condition for the possibility of both. So whereas many interpreters of Aristotle read *proteron* and *husteron* as non-temporal, putting place first, Heidegger argues that "the experience of before and after intrinsically presupposes, in a certain way, the earlier and later" (BPPb 247/349). On his interpretation, Aristotle does not reduce time to space but brings to light a phenomenon of time that is prior to both. The ingenuity of Heidegger's claim lies in the way he credits Aristotle with catching sight of the difference between time and temporality, and its audacity lies in his denial of the primacy of place explicitly affirmed by Aristotle.

To show how Aristotle sheds light not only on the horizonal structure of temporality but also on its ecstatic character, Heidegger performs another reversal. In his interpretation, Aristotle does not correlate nows with spatial points, but clarifies ecstatic temporality by describing the "peculiar double visage" of the now as it stretches out toward the future and the past. Once again, however, Heidegger reverses the order of priority established by Aristotle, in this case between the now as limit and the now as transition. While Aristotle does say that the now, when functioning as a boundary, "is not time, but an attribute of it,"[31] he later argues in *Physics* VI that the now must be an indivisible point between the past and future.[32] Aristotle worries that if the now were divisible or extended—if it stretched even imperceptibly—then problems would arise with the past, present, and future overlapping one another. Only by ignoring these arguments can Heidegger maintain that the Aristotelian now is always already "beyond the point" and that Bergson misunderstands Aristotle by overlooking this.

Finally, Heidegger insists that the metaphor of time as a container should also be interpreted as a formal, ontological structure. He cautions that Aristotle's descriptions of time and number "containing" things that can be measured by them should not be taken literally, and he interprets this characterization as more fundamental than spatial location. As with other terms he retranslates to mark their status as formal indications, Heidegger speaks of things "*being embraced* by time." Thus, Heidegger reads Aristotle as clarifying the meaning of intratemporality without reference to spatial relations: being "in" time is analogous to being "in" number. But again, as in his discussion of how time "follows" motion and extension, Aristotle says nothing to discourage thinking in terms of space, and he even concludes his remarks on time and number: "So it is necessary that all the things in time should be contained by time, just like other things also which are in anything, e.g., the things in place by place."[33] Ignoring this remark, Heidegger focuses instead on how Aristotle's explanation of intratemporality clarifies the traditional function of time as a criterion for distinguishing entities that undergo change, and can be measured accordingly, from those that are neither in motion nor at rest. On this reading, it is number and measurability that are signified by the "in," not place. Rather than reversing Aristotle's order of priority in this case, Heidegger simply leaves out his reference to place.

Heidegger's positive appropriation of Aristotle in *Basic Problems* presents him as not just defining time on the basis of the ordinary understanding of it, but also catching a glimpse of the structure of temporality. In Aristotle's discussion of the now, he reveals its ecstatic openness to the past and the future. And in his discussion of time and number, Aristotle reveals the character of time as an enclosing horizon, always both earlier and later than the things contained in it. What is so interesting about this with respect to Heidegger's critique of Bergson, though, is that Heidegger does this precisely by reading Aristotle as if he had anticipated the Bergsonian worry about reducing time to space. In effect, Heidegger uncovers hidden possibilities in Aristotle's account of time by de-spatializing his thinking, that is, by showing how he thinks time in terms of temporality rather than space. However, a closer look the misunderstandings that Heidegger attributes to Bergson has shown that he not only exaggerates the extent to which Bergson is Aristotelian, but he also portrays Aristotle as avoiding any reduction of time to space, and in that respect—as anachronistic as it may sound—as being Bergsonian.

6. An Inverted History of Time

Given that *Basic Problems* represents an attempt by Heidegger to go further with the project of *Being and Time*, and given the way he associates Bergson with Aristotle, it makes sense that he would use his explication of *Physics* IV as an opportunity to confront Bergson. What I have shown, however, is that Heidegger's continued engagement with Bergson is crucial for his positive appropriation of Aristotle's concept of time. First, it is clear that Heidegger's interpretation of Aristotle is shaped by his opposition to Bergson. To be precise, Heidegger uses the complaint, attributed to Bergson, that "time is space" for Aristotle to develop a reading that stands in contrast to the critical remarks in *Being and Time* focusing on how Aristotelian now-time levels down the full structure of temporality. In his second reading, Heidegger emphasizes not the succession of nows that characterizes the traditional concept of time, but the "determinations of transition, holding-around, and unveiledness" (BPPb 273/387) that are hidden in Aristotle's account of time. On this reading, Aristotle catches sight of the structure of ecstatic-horizonal temporality, deriving time not from space, but from a more primordial phenomenon of time. Thus, rather than trying to demonstrate how Bergson's way of thinking about time is determined by his reading of Aristotle, Heidegger takes this for granted and presents Aristotle as if he had already anticipated Bergson's concern. This provides Heidegger with the leverage he needs to lift Aristotle above standard interpretations of his account of time, at once liberating him from the tradition he inaugurated and confining Bergson within it.

Not only does Heidegger portray Bergson as Aristotelian with respect to time, but, by arguing that Aristotle never thinks of time in terms of space, he also portrays Aristotle as Bergsonian *avant la lettre*. This is because Heidegger's strategy of complementing his critical interpretation of Aristotle with a positive appropriation of his insight into temporality relies on a vigilant de-spatialization of Aristotle's thinking about time. To this end, Heidegger ignores crucial passages, reverses orders of priority, and reads ontological structures into Aristotle's text in order to meet Bergson's challenge not to think about time in terms of space. The result is an inverted world, or an inverted history of the concept of time, in which Aristotle offers a radical phenomenology of time while Bergson's way of thinking is thoroughly traditional. But, of course, this conclusion is based on only one of Heidegger's readings of Aristotle. On the more critical reading, Aristotelian "now-time" is precisely what levels down originary temporality by privileging presence and treating time as one being among others.

When we recall that Heidegger offers no real support for the claim that Bergson accuses Aristotle of reducing time to space, and it becomes clear that he corrects Bergson's alleged misunderstandings with a reading of Aristotle that could justifiably be called Bergsonian, Heidegger's dismissal of Bergson becomes even more questionable. The notion that Bergson's way of thinking about time is merely a reversal of Aristotle has turned out to be a convenient fiction, one that Heidegger uses to his advantage in showing how originary temporality is there from the beginning. In effect, Heidegger uses an oversimplified Bergsonism to de-spatialize Aristotle's concept of time and uncover the untapped resources of his thinking. Heidegger dismisses Bergson in *Being and Time* for merely reversing Aristotle's concept of time, but it is actually Heidegger who reverses Bergsonism. By subordinating Bergson's thinking to Aristotle's, Heidegger attempts to show that even Bergson operates within the confines of Greek ontology, equating being with presence. Because Bergson develops the concept of duration in response to the confusion of time with space, which supposedly underlies the concept of time as something counted in connection with motion, he does not get to the root of the problem. Bergson's critique misses the point, for Heidegger, because it does not challenge the privilege of presence. Yet Heidegger appears to realize that originary temporality can be discerned in Aristotle's thinking by showing how he avoids confusing time with space.

Finally, there is another problem with Heidegger's critique of Bergson that we have yet to consider. Although Heidegger acknowledges the importance of Bergson's other works, his interpretation stays focused on the concept of duration developed in *Time and Free Will.* Yet in Bergson's second major work, *Matter and Memory*, he advances his thinking in a number of new directions, broadening his concept of duration beyond consciousness and even calling the being of consciousness into question. With his theory of pure memory, Bergson even questions the very feature of traditional ontology that Heidegger identifies as the root of the problem with time: the interpretation of being as presence. By looking more closely at the ways Bergson's understanding of duration changes in *Matter and Memory*, we can see how much of what is radical in his thinking Heidegger ignores.

5

Challenging the Privilege of Presence

The Ontological Turn in *Matter and Memory*

1. From Duration to Memory

In the course of Heidegger's reading of Aristotle in *The Basic Problems of Phenomenology*, he attempts to show not only that Bergson thinks about time in a fundamentally Aristotelian way, as he had claimed in *Being and Time*, but also that Bergson's interpretation of time "rests on a misunderstanding of Aristotle's way of understanding time" (BPP 232/329). More specifically, Heidegger argues that Bergson's opposition to Aristotle is based on the misconception that Aristotle reduces time to space. On the one hand, he claims that Bergsonian duration is a mere reversal of Aristotle's concept of time, and on the other hand, that Bergson's attack on time as a spatial representation is based on a misinterpretation of that concept. I have raised some problems with both lines of criticism and called into question Heidegger's dismissal of Bergson as well as his reading of Aristotle. Now, even though Heidegger takes certain liberties in claiming that Bergson's thinking is fundamentally Aristotelian, and even though we have good reasons to question Heidegger's account of what Bergson misunderstands about Aristotle, we still have to come to terms with two of his basic concerns about Bergson's philosophy of time.

The first concern is evident in Heidegger's association of Bergson early in *Being and Time* with Dilthey's interpretation of life, Scheler's analysis of the person, and Husserl's idea of pure consciousness. The concern is that Bergson, like the contemporary thinkers with whom Heidegger is largely sympathetic, *neglects the question of being*. This concern underlies Heidegger's pregnant remark that the only advances beyond Kant and Aristotle in Bergson's thinking have to do with "grasping time and 'time consciousness'" (BT 410n/433n). The language Heidegger uses suggests that he views

Bergson's understanding of time as sharing something with Husserl's, namely an orientation toward the internal, conscious apprehension of time.[1] The basic problem with this, as he argued in *History of the Concept of Time*, is that consciousness as a "thematic field" of investigation needs ontological clarification. The *being of consciousness* is immanence, for Husserl, and he understands consciousness as a field of pure subjectivity, but he does not adequately address the question of its being. Heidegger maintains that it is no use to distinguish the immanent or subjective time of consciousness, or lived experience, from the transcendent or objective time of the world, as long as the former is just a reversal of the latter.[2] In his view, the primordial phenomenon of time remains hidden from all investigations into time-consciousness, whether phenomenological like Husserl's or psychological like Bergson's and James's. If Bergson takes consciousness as his field of investigation and maintains a traditional view of what consciousness is, then this blocks his access to originary temporality.

Heidegger's second concern is evident in his association of Bergson with not only Aristotle but also Hegel, who understands time in terms of the "now." Heidegger worries that Bergson's ontological neglect prevents him from recognizing how the traditional concept of time as a succession of nows is based on the assumption that *being equals presence*, or that whatever exists must be present in some way. Aristotle and the entire ontological tradition treat time as something objectively present or else nonexistent, on Heidegger's account. This explains the primacy of the now in the interpretations of time offered by even the most apparently radical (and radically opposed) thinkers, including Hegel and Bergson. For Heidegger, Bergson remains traditional in the way he understands time because he *fails to challenge the privilege of presence*. Again, this failure prevents him from bringing originary temporality to light.

Heidegger's critique of Bergson in *Being and Time* focuses on ideas that appear in *Time and Free Will*, a significant choice because it is an early work that stands at the beginning of a career spanning four and a half decades. Heidegger recognized that to take Bergson's thought seriously, he would eventually have to deal with *Matter and Memory* as well. In the 1925–6 *Logic* course in which Heidegger first lays out his critique of Bergson, he remarks:

> In his later writings Bergson has not changed the exposition of time given in his early work. On the contrary, he has maintained it up to the present day. The essential and enduring element of his philosophical work does not lie in this direction at all. Rather, the things of value, for which we are grateful, are found in his

> text *Matière et Mémoire*. It is a basic text for modern biology, and it contains insights that are far from exhausted yet. (LQT 207–8/250–1)

These comments raise a number of questions. Why would Heidegger continue, in *Being and Time* and beyond, to attack aspects of Bergson's thought that he did not consider essential? What does Heidegger find so valuable in *Matter and Memory*? Why does he characterize it as a contribution to biology as opposed to psychology or metaphysics, the fields Bergson addresses most directly?[3]

In my view, Heidegger is correct to stress the importance of *Matter and Memory* for the reason that Bergson's thinking about time *does* change as he turns his attention from time to memory. In fact, Bergson's concept of duration undergoes an evolution in *Matter and Memory* that Heidegger ignores.[4] Without abandoning the view developed in *Time and Free Will*, Bergson attempts to overcome the dualism of that work by rethinking consciousness and its relationship to duration. In *Time and Free Will*, Bergson characterizes duration as the lived experience of the passage of time, a flux whose moments permeate one another rather than being juxtaposed like objects in space. It is, in Bergson's terms, the continuous and heterogeneous multiplicity of conscious states, as opposed to the discrete multiplicity of objects in space or numbers in a series. Duration depends on memory, which synthesizes successive moments into an organic whole, and just as there is no extensity in consciousness, there is no duration in space. Thus, in *Time and Free Will*, duration appears to be a psychological experience, and Bergson goes so far as to argue that objects only endure for conscious observers who project duration into the world. In *Matter and Memory*, however, Bergson takes an ontological turn that complicates his understanding of duration in crucial ways that Heidegger's critique does not take into account. The details of this turn in Bergson's thinking are the focus of my attempt to address Heidegger's concerns.[5]

To see how Bergson's thinking evolves in *Matter and Memory*, it is necessary to understand how that work provides a response to the mind-body problem, another great problem of metaphysics, by means of a temporal interpretation of the phenomena involved.[6] Bergson hopes to show, again by rejecting certain assumptions and habits of thinking at the source of the problem, that "*[q]uestions relating to subject and object, to their distinction and their union, should be put in terms of time rather than of space*" (MM 71/218). In so doing, he develops a theory of *perception* as the discernment of bodily action and a theory of *memory* as the movement of mind or spirit (*l'esprit*). If mind and body, consciousness and matter, can be understood in terms of time, then there is no need to reduce one to the other (as in materialism and idealism) or to view them as fundamentally opposed (as in ordinary dualism). The attempt to understand mind and body in terms of time leads Bergson to characterize perception and matter, both being present, as different in degree only, but perception and memory as different in kind. Bergson argues that while actual, concrete perception is a mixture of the two, they should not be confused; there is a fundamental temporal difference between perception and memory. Moreover, despite the common notion that the past has ceased to exist, he claims that it must continue to exist without being present to consciousness or stored in the brain. The difference between present objects and memories of the past is, he argues, a practical distinction that has been wrongly elevated to a metaphysical one. If mind and matter can both be understood in terms of duration, then we can avoid reducing one to the other and achieve a better understanding of their interaction. But duration can no longer be regarded as an inner experience, a feature of consciousness projected into the world. Duration becomes, for Bergson, a way to think of the difference between ourselves and other beings, as well as our openness to them. Thus, in *Matter and Memory*, contrary to Heidegger's critique, Bergson actually does call the being of consciousness and the privilege of presence into question.[7]

2. The Presence of Images

If, in *Time and Free Will*, Bergson maintains a dualism between consciousness—unextended, qualitative, and free—and the external world—extended, quantitative, and determined—in *Matter and Memory* he aims to "throw some light on the reciprocal action of spirit and matter" by "suppressing or toning down these three oppositions" (MM 244/373).[8] His strategy is to show, as he did with the free will debate, that the opposing sides—realism (or materialism) and idealism—share common assumptions and habits of thinking that are the source of the debate over the relationship between mind and body. He begins with an assumption common to both sides of the metaphysical debate, namely that "*perception has a wholly speculative interest*," or, in other words, "to perceive means above all to know" (MM 28/179). This epistemological prejudice is responsible for the view of perception as a subjective, interior representation of things, which Bergson rejects. Realists and idealists alike thus think of perception and matter as "different in nature" or "different in kind," and they attempt to overcome dualism by reducing mind to matter or matter to mind. They therefore make the crucial mistake of thinking that there is "only a difference of intensity instead of a difference of nature, between pure perception and memory" (MM 67/215). If we suppose that perception is directed primarily toward *action*, rather than knowledge, Bergson argues, then we will see that the difference between it and memory is more fundamental than the difference between it and matter. Consciousness, understood in terms of action rather than knowledge, will turn out to be subjective by virtue of the contributions of memory and affection, but also objective by virtue of the contributions of matter.[9]

Bergson attempts to redraw the battle lines in the mind-body debate in the first chapter of *Matter and Memory* through a theory of "pure perception."[10] He begins with a quasi-phenomenological suspension of metaphysical theories and a return to a pre-theoretical standpoint characterized by a certain ambiguity: "Here I am in the presence of images, in the vaguest sense of the word, images perceived when my senses are open to them, unperceived when they are closed" (MM 17/169). On the one hand, these images manifest something objective and calculable, a world in which "the future of the images must be contained in the present" (MM 17/169). On the other hand, they are pervaded by subjectivity, thanks to being coupled with feelings and sensations (which Bergson now calls "affections"), and by unpredictability, because "the act in which an affective state issues is not one of those which might be rigorously deduced from antecedent phenomena . . . hence, it really adds something new to the universe and to history" (MM 18/170). The presence of images is thus defined by a dual necessity: that every image acts reciprocally and according to the same laws, and that these images surround a privileged image, "my body," which not only reacts mechanically but also has the capacity to act by choice. The images we perceive through our senses thus belong simultaneously to "two distinct systems," according to Bergson: the order of nature, which he identifies with science, and "*consciousness*, wherein all the images depend on a central image, our body, the variations of which they follow" (MM 26/177). The difference between realism and idealism is that one privileges the standpoint of science and attempts to reduce mind to matter, while the other privileges the standpoint of consciousness and attempts to reduce matter to mind. To reveal the common assumptions at the root of the mind-body problem, Bergson proposes that we think of the world as an "aggregate of 'images,'" where an image is not a mere mental state, but "an existence situated halfway between the 'thing' and the 'representation' " (MM 9/161). The question then becomes neither how do material objects produce ideas in the mind nor how does the mind constitute an external world, but how do images belong to both systems at once?[11]

While Bergson's most radical challenge to traditional ontology occurs with his theory of pure memory, his theory of pure perception sets the stage by considering what perception would be, in principle, if there were no memory involved. Suspending the habit of thinking about perception in terms of knowledge, Bergson considers the part that conscious perception plays in the reactions of a living body to its environment. That is, he interprets perception as first and foremost embodied.[12] In the totality of images constituting the material world, Bergson says, one's own body is something grasped "from without by perceptions, but from within by affections" (MM 17/169). This body is part of the world, and as it both moves and is moved by the images in its immediate environment—receiving stimuli and responding—it is a "center of action" (MM 20/172). Understood in terms of action, the primary role of perception is not to produce representations in the mind, but to support the body in executing movements and navigating the world.

Perception is not always required for action, in Bergson's account, but it broadens the scope of possible action. In reflexive activity, for example, stimuli produce an immediate, automatic response, and conscious perception is bypassed. However, when the living body faces the task of having to *choose* an appropriate response, "perception appears at the precise moment when a stimulation received by matter is not prolonged into a necessary action" (MM 32/182). Perception is thus a function of indetermination in bodies capable of voluntary activity, for Bergson. From simple, single-celled organisms who depend on touch to identify "promises and threats" in their environment, to higher vertebrates whose sight and hearing vastly expands their range of possible actions, perception is "a *variable* relation between the living being and the more-or-less distant influence of the objects which interest it" (MM 33/183). More specifically, perception varies with the sophistication the body's nervous system: "The more it develops, the more numerous and distant are the points of space which it brings into relation with ever more complex motor mechanisms. In this way the scope which it allows to our action enlarges" (MM 31/181). If conscious perception is anything like knowledge, it is because it is the body's way of acquainting itself with or discovering what it is capable of doing in its present situation.[13]

On the basis of this view of perception as directed primarily toward action, Bergson challenges the notion of consciousness as producing or reproducing external things. Contrary to how perception is often understood, he claims, our brains or bodies "add nothing to what is there" (MM 39/188). That is, perception is neither the production of extended objects out of unextended sensations, as idealists tend to think, nor the reproduction of objects in the form of mental representations, as realists do. Pure perception is, for Bergson, a diminution of matter rather than an addition to or "spiritualization" of it. To clarify this, Bergson reverses the metaphor of consciousness illuminating objects, which suggests that perception adds something to that which is perceived. According to his opening hypothesis, an image in the material world is bound by necessity "to transmit the whole of what it receives, to oppose to every action an equal and contrary reaction" (MM 36/186). The image of one's own body, on the other hand, has a capacity for choice that, while not fundamentally different from reflex action, involves the capacity to react to some stimuli and not others, to act immediately or to consider a variety of possible actions before responding. At the physiological level, Bergson rejects the notion that the brain is an organ capable of producing representations of things; instead, what the brain can do is delay the transmission of stimuli in order to select an appropriate action. In the interval between stimulus and response, perception subtracts from or diminishes images according to our interests. In reflecting the aspects of objects that are relevant to our possible action, the role of perception is "not to throw more light on the object, but, on the contrary, to obscure some of its aspects" (MM 36/186). Thus, for Bergson, we should reverse the metaphor of consciousness as illuminating objects and think of it instead as screening out things that are uninteresting for the sake of possible action.[14]

Contrary to idealism and materialism, Bergson maintains, conscious perception neither produces nor reproduces images but works as a filter. In the "zones of indetermination" surrounding the body, Bergson says, "the real action passes through, the virtual action remains" (MM 39/189). It is only by screening out the uninteresting or useless aspects of objects that the living body is capable of voluntary action. According to the epistemological prejudice that Bergson rejects, complete perception would be perfect knowledge. However, in terms of action, "[t]o perceive all the influences from all the points of all bodies would be to descend to the condition of a material object. Conscious perception signifies choice, and consciousness mainly consists in this practical discernment" (MM 49/198). Consciousness is *discernment* in the sense that it distinguishes or separates what is relevant to our vital needs and interests from what is not. Perception thus does not produce disinterested, disembodied representations of material objects, but rather screens out what has no interest for bodily functions.

Bergson's view of consciousness as a filter helps to explain how it can be that perception and matter are not different in kind. If matter is, as he has hypothesized, "the aggregate of images," then perception can be understood as "these same images referred to the eventual action of one particular image, my body" (MM 22/173). Bergson has shown how the same images can belong to two different systems, nature and consciousness: in the former, images appear in the light of the movement of every other image in the universe, while in the latter, images reflect our needs and interests, the virtual and yet-to-be-determined action of "my body." As a precursor to real action performed by the body, "consciousness means virtual action" (MM 50/199). As such, Bergson denies that perception is subjective, internal, and unextended. Instead of taking place in the mind or even the brain, he concludes, "[p]erception, in its pure state, is, then, in very truth, a part of things" (MM 64/212). If we must think in spatial terms, we should say that *perception takes place in the things perceived*, thus that it is objective, external, and impersonal.[15] In temporal terms, perception and matter are not fundamentally different because they both occur in the present and only endure thanks to memory.

It is crucial for Bergson that perceptual consciousness is not an enclosed sphere of subjectivity. He briefly but explicitly expresses his view in ontological terms when he considers what distinguishes a material object from a representation. He writes:

> It is true that an image may *be* without *being perceived*—it may be present without being represented—and the distance between these two terms, presence and representation, seems just to measure the interval between matter itself and our conscious perception of matter. (MM 35/185)

Here Bergson seems to equate being with being present, the mark of traditional ontology for Heidegger. But he is trying to articulate how material images become perceptual images—in other words, how presence is related to representation—without taking memory into account yet. At this stage, his primary concern is to justify this conclusion:

> There is for images merely a difference of degree, and not of kind, between *being* and *being consciously perceived*. The reality of matter consists in the totality of its elements and of their actions of every kind. Our representation of matter is the measure of our possible action upon bodies: it results from the discarding of what has no interest for our needs, or more generally, for our functions. (MM 37–38/187–8)

An object and a representation are thus two different modes of an image, one in absolute reciprocity with every other image, the other appearing to consciousness relative to the indetermination of our bodily movements. The representation of a thing is characterized by the "necessary poverty" of our practical discernment, which presents objects not in every detail, but in outlines or contours corresponding to our interests. For Bergson, the traditional opposition between being and being consciously perceived must be rethought, beginning with an acknowledgement that the subjectivity and even the interiority of consciousness comes not from perception but from memory.

In his analysis of perception, Bergson may not give ontology the absolute priority that Heidegger does, but neither does he shy away from it. In fact, his theory of pure perception raises the question of the being of consciousness in a precise and provocative way. When Bergson first presents consciousness as a system of images oriented around the body as a center of action, he departs from the usual practice of regarding consciousness as purely psychological, distinct from anything physical. Instead, he challenges the traditional notion of consciousness as disembodied, separate and distinct from the world, and limited to human beings. Consciousness begins to appear in much simpler organisms, ones whose actions are not all automatic responses to the things in their environment. Bergson interprets consciousness in terms of conscious perception and presents it as a function of indetermination, or the ability of a body to choose from among a range of possible actions. On his account, there is no conscious perception without hesitation or delay in the body's execution of movements; consciousness occurs in the interval between action and reaction, and it is the virtual action of the body reflected by the objects around it.[16] Conscious perception is therefore not a disinterested, internal representation of external things, for Bergson; it is practical discernment, which takes place in the world, in the things perceived, rather than in the mind, as realists, idealists, and ordinary dualists have assumed. Consciousness neither reproduces objects nor produces them in the first place, but screens out the aspects that have no vital interest for us. Contrary to the traditional view, Bergson maintains that perception is external, objective, and impersonal; it is related to matter like a part to the whole. It follows that being consciously perceived, or perception, is no different in kind from being, or matter. In this way, by suspending the habit of thinking of perception in terms of knowledge, Bergson directly challenges the view of consciousness as a subjective, inner realm of representations.

3. The Survival of the Past in the Present: Memory and Habit

While Bergson maintains that perception is no different in kind from matter, he goes on to emphasize that it is only one aspect of consciousness. For pure perception, unmixed with memory, would be instantaneous; it would have no duration and therefore would be cut off from every other moment of time. In reality, though, perception "always occupies a certain duration, and involves, consequently, an effort of memory" (MM 34/184). Actual consciousness is not an "instantaneous vision" of the world, but a series of moments joined together on the "continuous thread of memory" (MM 69/216). Bergson's theory of pure perception has only dealt with the "objective side" of consciousness, its contact or coincidence with matter, in order to be able to show more clearly what constitutes its "subjective side." While it is common to assume that perception is "a kind of *interior* and *subjective* vision, which would then differ from memory only by its greater intensity" (MM 34/184), Bergson argues that it is only subjective insofar as it is mixed with affection, which occurs inside the body, and memory, which brings our past experience to bear on the present. The view of consciousness offered so far in *Matter and Memory* thus differs greatly from the one given in *Time and Free Will* because Bergson has left out precisely that which makes it a continuous or qualitative multiplicity. Now he explains: "The qualitative heterogeneity of our successive perceptions of the universe results from the fact that each, in itself, extends over a certain depth of duration and that memory condenses in each an enormous multiplicity of vibrations which appear to us all at once, although they are successive" (MM 70/217). Apart from memory, perception accounts for how images of things can be present to consciousness, but not how they can be connected in a continuous, coherent experience. There would be no consciousness of the sort presupposed by philosophers on both sides of the mind-body debate, and no subjectivity, were it not for memory.

It is at this point that Bergson proposes that the mind-body problem, and all questions relating to subject and object, be rethought in terms of time, in contrast to how it has traditionally been understood in terms of space. Bergson maintains that even though all perception is mixed with memory—so "pure perception" is not actual but ideal—there is a fundamental difference, a "difference in kind," between perception and memory. While perception and recollection "always interpenetrate one another," he claims, "if we make recollection [*souvenir*] merely a weakened perception, we misunderstand the essential difference between the past and the present" (MM 67/214). Thus, Bergson attacks not only the assumption that perception is directed primarily toward knowledge, but also the tendency to think of memories as nothing but weaker or less intense perceptions. The difference between perception and memory is not one of vitality, in his view, but of *temporality*: perception in its pure form is directed toward the present (and action), while memory in its pure form is directed toward the past (and recollection). For Bergson, focusing on the temporal distinction will make it possible to understand memory as "a power absolutely independent of matter" and as "spirit in its most tangible form" (MM 73/220). His aim is to show how the mind, understood temporally, both interacts with the body and is radically distinct from it.

There are many ways of dividing memory into types: episodic and semantic, voluntary and involuntary, explicit and implicit, retentive and recollective, and so on.[17] Bergson begins chapter two of *Matter and Memory* with a distinction of this sort: "*The past survives in two distinct forms: first, in motor mechanisms; secondly, in independent recollections*" (MM 78/224). We can see at a glance that these two forms of memory correspond to the distinction between body and mind. The memory of the body, for Bergson, "lies in the action itself and the automatic setting in motion of a mechanism adapted to the circumstances," while the memory of the mind "seeks in the past . . . representations which are best able to enter into the present situation" (MM 78/224). He illustrates the difference with the example of a lesson learned by rote memorization. When we recite a passage or play a piece of music from memory, he says, the recitation has "*all* the marks of a habit" (MM 80/225): it is a skill, a series of coordinated movements that was acquired gradually by repetition, and it can be performed automatically, without thought or effort. In contrast, when we remember reading or playing the same piece on a particular occasion, the memory has "*none* of the marks of a habit": it is an image that was imprinted immediately, without repetition; it is unique, "like an event in my life," located in the past along with surrounding events, thus "unable to occur again"; and it is "perfect," that is, unalterable except through forgetting, "necessarily at the outset what it will always be" (MM 80/225). Bergson thus summarizes the "profound difference" between these two forms of memory in this way: while the latter is a *representation*, which we can grasp in an instant or take time to recall in detail, the former is an *action*, which we need a definite time to perform. Here, then, we have two "theoretically independent" forms of memory (*mémoire*), one that preserves past events and recalls them in the form of memory-images (*images-souvenirs*) and another that "prolongs" the past into the present through movements performed mechanically by the body.[18]

Bergson maintains that these two forms of memory correspond to two fundamentally different ways that the past "survives" in the present. On the one hand, the past survives in *images* that are spontaneously recalled. Memory-images include a wealth of detail about past events, including their order with respect to other events. We can recall an event—the recital of a poem or a piece of music, for example—voluntarily, or the image may return to us automatically. Such "spontaneous recollection" may be like a photograph that condenses an event into an instant, or it may be more like a film that we can speed up or slow down. In either case, "time can add nothing to its image without disfiguring it; it retains in memory its place and date" (MM 83/229). Bergson calls this form of memory "regressive" to indicate that it is primarily backward-directed, turning away from action and life. He also calls it "true memory" and "memory *par excellence*," as opposed to the bodily form, which is just "habit interpreted by memory" (MM 84/229).

While true memory represents the past, preserving it in detail "regardless of utility or of practical application," bodily memory merely repeats it, remaining "bent upon action, seated in the present and looking only to the future" (MM 81–82/227). This repetition of past *movements*, as opposed to images, is forward-directed, we might even say progressive.[19] Thanks to it, the past survives not only in images, but also in "motor mechanisms" and "motor tendencies" of the body. Activities like reciting a poem or playing a piece of music involve repeating the same movements over and over until they can be performed automatically. Bodily memory thus involves the repetition of effort, the "decomposition and then recomposition" of action, and the production of "a mechanism which is set in motion as a whole by an initial impulse, in a closed system of automatic movements which succeed each other in the same order and, together, take the same length of time" (MM 80/226). In this case, it is not images of the past that are preserved, but movements "deposited" in the body, and recollection is not spontaneous, but learned. Rather than being preserved in the past with its "place and date" intact, Bergson suggests, "a learned recollection passes out of time in the measure that the lesson is better known; it becomes more and more impersonal, more and more foreign to our past life" (MM 83/229). Bodily memory is thus geared toward possible action, prolonging the past into the present not by retaining impressions, but by repeating movements.[20]

What we ordinarily think of as memory is actually, for Bergson, a mixture of bodily memory and true memory, which are often confused. To show even more clearly how they are distinct, Bergson considers their relationship to each other. When memorizing something or acquiring a habit, we make use of true memory, especially in the early stages, to coordinate our movements into a structured routine. Learning a new piece of music, for instance, may require us to recall which note is supposed to be played next, where each finger should land, and so forth. In this case, "the two memories run side by side and lend to each other a mutual support" (MM 86/231). However, the goal of learning a piece is to be able to play it without having to remember what to do next. Although it takes effort to learn, once we know a piece by heart, we play it more or less automatically, so that remembering a particular practice session would serve no purpose and could even be disruptive. For Bergson, this illustrates how regressive memory requires us "to withdraw ourselves from the action of the moment" and "to value the useless" (MM 83/228). At the same time, our practical needs ordinarily prevent us from dwelling in the past, that is, experiencing "useless" memory-images, which have no similarity to the images being perceived and can contribute nothing to our action. Because acquired habits or bodily mechanisms are artifices that enable us to "prolong action" by repeating the past without having to represent it, bodily memory usually *inhibits* true memory, "or at least will only accept from it that which can throw light upon and complete the present situation" (MM 85/230). So whereas the two forms of memory work together when new motor mechanisms are being learned, later, for the sake of utility, those mechanisms of bodily memory block memory-images from intruding on the present.

The experience of recognition further illustrates the difference between these two forms of memory, and Bergson draws a parallel distinction between inattentive and attentive recognition. In opposition to the associationist theory of mind, which tends to overlook the role of habit, he claims: "*The recognition of a present object is effected by movements when it proceeds from the object, by representations when it issues from the subject*" (MM 78/224). According to associationism, when we recognize an object, a face, a word, or a place, our perception of it brings back memories of having perceived it in the past. On the contrary, Bergson argues, recognition usually does not require memory-images, but only the adoption of a "bodily attitude" or "motor tendency" triggered by perception. To illustrate this, he describes the experience of navigating an unfamiliar city in contrast to taking a walk close to home. In a new city, we move haltingly, block by block, hesitating each time we need to make a choice about which direction to go, as opposed to how we move through our own neighborhood "mechanically, without having any distinct perception of the objects which [we are] passing . . . hardly conscious of anything but automatism" (MM 93/238). Our bodies become habituated to familiar routes, and like our ability to recite something by heart, our recognition of the streets, buildings, and landmarks is based on motor memory. This type of recognition depends on movements rather than images, and our feeling of recognition is the conscious counterpart to bodily attitudes adopted more or less automatically. Not only are memory-images unnecessary for this kind of recognition, but they also need to be inhibited by the motor mechanisms that make it possible. Like a gatekeeper, our well-coordinated movements turn away memories that have no bearing on the present, but they welcome those from similar situations in the past that help to sustain our action.

Bergson classifies the kind of recognition experienced when walking through one's own neighborhood as "inattentive" because it does not require the same effort of conscious perception as getting around in an unfamiliar place. He distinguishes this from the kind of recognition that does not happen automatically, but requires us to "search our memory." This *attentive* recognition is based on images rather than movements, but even it begins with bodily attitudes and movements, according to Bergson. The distinction between attentive and inattentive recognition parallels the distinction between true memory, which employs memory-images to represent the past, and bodily memory, which employs motor mechanisms to repeat it. Attentive recognition "strengthens and enriches" perception with memories of similar experiences in the past. Inattentive or automatic recognition occurs when there is no need to arrest movement and wait for memory-images because the "organized motor accompaniment" (MM 94/240) to perception is sufficient. Bergson describes these forms of recognition as following two opposite tendencies, some that "*take us away* from the object" by means of motor mechanisms—the inattentive—and others that "*bring us back* to the object" by means of memory-images (MM 98/244)—the attentive. Memory thus plays a leading role in our ability not just to recognize the things around us, but also to linger over them and perceive them in greater detail.

In fact, the phenomenon of attention clarifies the roles that both forms of memory play in perception. In Bergson's view, conscious perception implies an interval between stimulus and response, a *delay* between sensations received by the body and motor mechanisms engaged to perform an action, which makes choice possible. This delay or hesitation reveals the "negative condition" of attention, namely that bodily movement be arrested in order to focus the mind on an object. But what accounts for our ability to perceive an object in greater and greater detail without it (or its surroundings) changing? Bergson's explanation is that attention depends on a "circuit" of perception and memory through which a perceptual image is fleshed out by memory-images. When we study an object to perceive it in great detail, not only can we move around it to take different points of view, but we can also stop moving and linger over it. When we do so, Bergson says, we arrest bodily movements directed to utility, and "more subtle movements graft themselves . . . [and] retrace the outlines of the object perceived" (MM 101/247). As perception "retraces" the contours of the object sketched out by habit, true memory retrieves images "which resemble it and which are already sketched out by the movement itself" (MM 101/247). Then the motor mechanisms at the basis of recognition are suspended, and their inhibition of recollections from the past is relaxed, allowing memory-images to fill in the contours of the perceptual image. As we continue to attend to the object, "an appeal is made to deeper and more distant regions of memory, until other details that are already known come to project themselves upon those details that remain unperceived. And the operation may go on indefinitely—memory strengthening and enriching perception" (MM 101/247). Perception becomes more attentive as we cycle through memory-images identical to the object perceived, ones similar to it, ones with a vague resemblance, and so on. But these images from the past do not stand out in the mind apart from the object; instead, they blend together in such a way that "any memory-image that is capable of interpreting our actual perception inserts itself so thoroughly into it that we are no longer able to discern what is perception and what is memory" (MM 103/248–9). The experience of reading offers a well-known illustration of how perception relies on memory to the extent that they become almost indistinguishable. Bergson notes that when we read, instead of seeing every letter of every word, we perceive just enough to be able to grasp the meaning, sometimes even projecting things onto the page that are not actually there. In this way, he says, "we are constantly creating or reconstructing. Our distinct perception is really comparable to a closed circle, in which the perception-image, going toward the mind, and the memory-image, launched into space, careen the one behind the other" (MM 103/249). Rather than a sequence of associated ideas, "reflective perception is a *circuit*, in which all the elements, including the perceived object itself, hold each other in a state of mutual tension" (MM 104/249). As Bergson will later explain, this tension is maintained by a constant back-and-forth movement between action and recollection, an oscillation that defines mental life as such.

Although Bergson argues that true memory and bodily memory should not be confused because they are different in kind, he nevertheless regards them as jointly responsible for our ability to perceive objects, navigate our surroundings, and even use language. For all such activities involve the exchange between movements and images involved in recognition. We can imagine a spectrum of cases, with some recognition happening automatically, as when we engage in a conversation in our native language, and others demanding an effort of recollection, as when we find ourselves lost in an unfamiliar place. In some cases, bodily movements, including those involved in speech, occur in such a fluid and well-coordinated manner that memory-images are not needed; in others, we have to wait for images from the past that afford us more ways to deal with the present situation. Bergson insists that such ordinary activities cannot be understood on the basis of the associationist theory of mind, for which "distinct perception and memory-image are taken in the static condition, as *things* of which the first is supposed to be already complete without the second; whereas we ought to consider the dynamic *progress* by which the one passes into the other" (MM 127/271). Indeed, for Bergson, the associationist tendency to think of all conscious states as ready-made objects is responsible for the confused notion that memories must be stored somewhere in the brain and that they can be destroyed by injury or disease. While the destruction of memories could account for cases of aphasia in which a patient cannot recognize familiar objects, places, or words, Bergson argues that injuries to the brain never destroy memories. Instead, they interrupt the circuit of perception and memory, preventing either the activation of motor mechanisms necessary for inattentive recognition or the recollection of memory-images necessary for attentive recognition.

Bergson's discussion of memory so far has focused on two ways the past survives: through the recollection of memory-images and the activation of motor mechanisms. He has called the former "true" memory, as opposed to mere habit, but he has not yet given an account of pure memory (*souvenir pur*). In fact, his discussion of recognition has focused on the process by which memory-images (*images souvenirs*) are "projected" or "grafted" onto perceptual images. In this process, memory blends with perception so completely, according to Bergson, that we cannot even tell where one begins and the other ends. Perceptual images are informed or filled in by recollection, and conversely, memory-images are manifested or embodied in the objects of perception. As such, memory-images are not "pure," but always mixed with perception. Because perception, unmixed with memory, would be completely present, Bergson asks, what would memory be if it were as completely past as perception is present?

4. The Survival of the Past in Itself

Bergson's theory of pure memory represents his most radical challenge to traditional ontology, as Heidegger views it, because it challenges the privilege of presence. Pure memory, or memory unmixed with perception, is implicit in Bergson's discussion of the two forms of memory, but he has yet to characterize it as such. He begins chapter three of *Matter and Memory* with a brief summary of the circuit he has described between the perception of a present object and the memory of past experiences, dividing it into three phases:

> *Perception* is never a mere contact of the mind with the object present; it is impregnated with memory-images which complete it as they interpret it. The *memory-image*, in its turn, partakes of "pure memory," which it begins to materialize, and of the perception in which it tends to embody itself. . . . Lastly, *pure memory*, though independent in theory, manifests itself as a rule only in the colored and living image which reveals it. (MM 133/276, emphasis added)

The first two phases of this process have been the subjects of the first two chapters of *Matter and Memory*, leading to the following conclusions: pure, instantaneous perception is the living body's discernment of matter according to its needs and interests, and memory-images are one of two ways that the past returns to the present (the other being motor mechanisms). While perceptual images sketch out present objects, memory-images revive the past. But to think of memory-images as dormant in the brain, waiting to be brought back by sensations, is to substitute a crude spatial image—"a discontinuity of multiple elements, inert and juxtaposed" (MM 134/277)—for a fundamentally temporal phenomenon.

The challenge, for Bergson, is to avoid spatial representations and think of memory in terms of time rather than space.[21] Whereas associationism tends to regard memories or recollections (*souvenirs*) as weaker, less vital copies of perceptions, Bergson argues that there is a more fundamental difference: perception is directed toward the present, and memory is directed toward the past. This is evident from the fact that we can always distinguish a memory from a sensation, even an extremely weak one, because memory is a representation of the past. A memory may give rise to a sensation, for example when we recall an embarrassing episode and relive the discomfort all over again. However, memory and sensation are always distinguished by an irreducible sense of "pastness" in the former. Memories are not just weaker or less intense than sensations, for Bergson, but they are temporally distinct: recollection is to the past as sensation and perception are to the present. To do justice to the difference in kind between memory and perception, Bergson aims to do something extraordinarily difficult: to think of pure memory, as opposed to the mixed phenomenon of the memory-image, as something that is in no way present.

In his view, pure memory cannot be understood merely as a collection of images lying dormant in the brain, because images are defined by presence. However, pure memory must have some role in the appearance of images that refer to past events. Bergson therefore interprets pure memory as a phase of the continuous process by which memory-images are presented to consciousness. Perception is always informed by memory, he claims, even when we are not conscious of what past experience contributes. It seems natural to assume that memories must be stored away somewhere in order to be accessed later, and if so, where but the brain could they be located? For Bergson, the question of *where memories exist* is a badly posed one that needs to be replaced by the question of *how the past survives in itself so that it can manifest itself in the present.* Associationism, assuming that memories must be present in some form, not only fails to see the difference in kind between memory and perception, but also confuses memory-images with pure memory. By regarding memory as a process or movement and pure memory as the immaterial phase, Bergson attempts to overcome the spatial metaphor of memories as dormant things, "inert and juxtaposed" elements, stored in the brain.

Because the difference in kind between perception and memory is based on the difference between the present and the past, Bergson offers an account of what distinguishes the past from the present. First and foremost, in his view, the difference between the present and the past is a practical one based on utility. In terms of action, he says, "My present is that which interests me, that which lives for me," whereas "my past is essentially powerless" (MM 137/280). In terms of duration, "time already flown is the past, and we call the present the instant where it flows" (MM 137/280, translation modified). However, the notion of the present as a "mathematical instant," as an "indivisible limit which separates the past from the future," is a merely symbolic substitute for "the real, concrete, lived present—that of which I speak when I speak of my present perception," which "necessarily occupies a duration" (MM 137/280). The "now" cannot be the basic phenomenon of time, for Bergson, because it is a mere abstraction. In actual perception, he writes, my present stretches out, with "one foot in my past and another in my future" (MM 138/280).[22] Even though the moment of pure perception has already passed as soon as I speak of it, it also "leans" or "points" in the direction of what is to come. In terms of the body, my present links the immediate past, in the form of sensations, with the immediate future, in the form of movement, and it therefore "consists in a joint system of sensations and movements" (MM 138/280–1). Thus, as a system of sensations and movements already described with regard to perception, as a "sensori-motor" phenomenon, "my present" refers to "the consciousness I have of my body" (MM 138/281).[23]

If the present consists of the set of sensations and movements belonging to our bodies, surrounded by the aggregate of images belonging to the material world, then any image belonging to that set is present rather than past. Without memory, Bergson says, my body "represents the actual state of my becoming, that part of my duration which is in process of growth" (MM 138/281). Matter exists in the moment, and can even be defined as "a present which is always beginning again" (MM 139/281). Conversely, the present, as a system of sensations and movements, is "the very materiality of our existence" (MM 139/281). This shows more clearly why perception and memory should be understood as different in kind: sensations "occupy definite portions of the surface of my body," while pure memory "interests no part of my body" (MM 139/281). Insofar as the present is *actual* and even material, recollection can be understood as the actualization or materialization of pure memory, which is *virtual*. Bergson says:

> From the moment that it becomes an image, the past leaves the state of pure memory and coincides with a certain part of my present. *Memory actualized in an image differs, then, profoundly from pure memory.* The image is a present state, and its sole share in the past is the memory from which it arose. Memory, on the contrary, powerless as long as it remains without utility, is pure from all admixture with sensation, is without attachment to the present, and is, consequently, unextended. (MM 140–1/283, emphasis added)

Memory-images, on this view, are an intermediate state between pure memory, which is "useless" and "powerless" as long as it has no bearing on action, and perception, which is concerned with utility and has the power to evoke past images.

This is a decisive point in Bergson's analysis. For pure memory to be limited as completely to the past as pure perception is to the present, it must be something profoundly different from the memory-images of which we are conscious. That is, pure memory cannot be contained in present consciousness. By defining conscious perception as a practical discernment of the aspects of matter relevant to the action of a living body, Bergson reversed the metaphor of consciousness as a spotlight: perception does not illuminate material objects so much as darken the aspects that do not interest us. Now he revives the old metaphor with respect to our choice of actions, claiming that consciousness "throws light on the immediate antecedents of the decision, and on those past recollections which can usefully combine with it; all else remains in shadow" (MM 141/283). If there is a spotlight of consciousness, he suggests, it illuminates the immediate past and events from the more remote past that are relevant to the present situation. The past is not "abolished" or "effaced" just because we are not conscious of it, according to Bergson. However, as most of the past remains in the dark, we must think of pure memory as something outside of consciousness or, to avoid spatial language, *unconscious*. Moreover, we must consider this: "How comes it then that an *existence outside of consciousness* appears clear to us in the case of objects, but obscure when we are speaking of the subject?" (MM 142/289). That is, why do we assume that distant, unperceived things exist but former states of consciousness do not?

With the acknowledgment that pure memory cannot be conscious, Bergson is ready to confront the ontological prejudice underlying the associationist theory of mind, as well as the debate between realism and idealism. He has already attacked the epistemological prejudice that perception is directed primarily toward knowledge. Now another assumption must be called into question: that *existence is limited to the present*. This prejudice manifests itself in realism and idealism in contradictory ways, Bergson argues, but for both, a practical difference between the present and the past is the hidden basis for the metaphysical distinction between existence and nonexistence. Interpreting the difference between the present and the past in terms of action, Bergson defines pure memory as the *virtual* existence of the past, thereby affirming the "survival of the past *in itself*" (MM 149/290) and offering a direct challenge to the privilege of presence. This challenge to traditional ontology makes the theory of pure memory one of the most radical and important aspects of Bergson's thought.

5. The Problem of Existence

The attempt to understand memory in terms of time rather than space has led Bergson to a crucial question: Why is it common to believe in the existence of objects in the material world that we do not currently perceive, but not in the existence of states of consciousness that we do not currently remember? Both are absent, but we regard them as having a different ontological status. On the one hand, few would claim that there is nothing beyond consciousness or that the images being perceived at this moment are the only things that exist. Instead, we ordinarily believe that there are such things as material objects, present to no one, existing outside of consciousness. On the other hand, we find it difficult to accept that psychological states experienced in the past—our sensations, feelings, thoughts, and so on—continue to exist after we are no longer conscious of them. Being in the past, such things *are* no longer—that is, they do not exist. So, Bergson notes, we are inclined to believe that *unperceived objects* exist, but not *unconscious representations*, even though neither are present. What is the basis of the supposed metaphysical difference between these two kinds of absence? Why do we tend to think of things in the past and the past itself as nonexistent? What exactly does it mean to exist?

For Bergson, our belief in the existence of unperceived objects but not past psychological states arises from the fact that the former are absent in space, while the latter are absent in time. Bergson explores the difference with the help of a simple diagram consisting of two intersecting lines: a horizontal line representing "all simultaneous objects in space" and a vertical line representing "our successive recollections set out in time" (MM 142–3/285). The axis stands for all that is present here and now, while the lines themselves stand for the two sets of absent objects, those in the world beyond our perception and those in our past experience. Why, he asks, are we inclined to attribute existence to things that are absent in space but to deny existence to those absent in time? To say that the former exist because they are in the world we perceive or are made of matter would be question-begging because we do not perceive the whole of matter, and it is the unperceived parts of the world that are at issue. According to Bergson, we can understand the difference if we make the effort to think about perception in terms of action. The set of objects in space includes things that are not now perceived but can be, while everything in the set of states in (past) time has already been perceived but cannot be perceived again. If something is incapable of being perceived, Bergson explains, then it is not a target for action. For practical purposes, "the past has no longer any interest for us; it has exhausted its possible action or will only recover an influence by borrowing the vitality of the present perception. The immediate future, on the contrary, consists in an impending action, in an energy not yet spent" (MM 144/285). As opposed to everything that has "flown by" or "passed through" our consciousness, the absent portion of the material world is full of invitations to act and is therefore useful and interesting. Thus, we consider objects in space, located at a distance from us, *actual*, even if we cannot perceive them, because they are still *actionable*, and we think that the past no longer exists because it cannot be acted upon. The distinction between absence in space and absence in time is thus, in Bergson's view, based entirely on utility.[24]

Going more deeply into this distinction, Bergson describes a double movement that, like the endosmosis between conscious states and external objects, is definitive for our thinking about both space and time. It is thanks to this movement that we view space as full of "objective realities without relation to consciousness" (MM 143/285) and time—specifically past time—as populated by states or events with no such reality. As Bergson showed in his analysis of perception, the farther our senses can reach into our surroundings, the longer we have for preparing to act, so "distance in space measures the proximity of a threat or of a promise in time" (MM 144/286). Here, the space to which Bergson refers is not the empty, homogeneous space of *Time and Free Will*, but the region surrounding a living body. In the perception of objects in our environment, he says, "space furnishes us at once with the diagram of our near future"—at least insofar as we perceive how long it will take for us to reach an object or for it to reach us—"and as this future must recede indefinitely, space which symbolizes it has for its property to remain, in its immobility, indefinitely open" (MM 144/286). That is, we always perceive objects in space against a horizon from which other objects and even other horizons can emerge. From a practical standpoint, "memories [*souvenirs*], on the contrary, inasmuch as they are past, are so much dead weight that we carry with us, and by which we prefer to imagine ourselves unencumbered" (MM 145/286). In terms of action, we are therefore inclined to regard the past as just as static or determined as the future is undetermined. Thus, the double movement, for Bergson: "The same instinct, in virtue of which we open out space indefinitely before us, prompts us to shut off time behind us as it flows" (MM 145/286). Because of this vital "instinct," the past is *closed*, and memories with no bearing on the present are "practically abolished," while space is *open* and even the most distant objects are thought of as existing. The closure of past time is, in Bergson's view, intrinsically connected to the opening of present space, just as the inhibition of useless memory-images is essential to the projection of useful ones. The practical distinction between what can be acted upon (in space) and what cannot (in time) thus gives rise to the appearance of the past as bygone, as "dead weight," Bergson contends, even though our entire past survives and animates the present. When we see that this distinction is rooted in our attention to life, he says, "the question is just whether the past has ceased to exist or whether it has simply ceased to be useful" (MM 149/291).[25]

According to Bergson's analysis, the abolition of the past and the metaphysical distinction between the existence of absent objects and the nonexistence of absent memories arises from a merely practical distinction, a difference of utility. He argues that our belief in the actuality of things in space but not of recollections in time stems from the habit of emphasizing the differences and ignoring the similarities between them. One difference is that the order of objects juxtaposed in space is necessary, while the order of states succeeding one another in time is contingent. That is, material objects are arranged and react to one another in predictable ways that memories, returning in an "apparently capricious" manner, do not. Our perception of surrounding objects is like a link in a chain, with the causal connection to other links bestowing an ontological status on absent objects that unconscious memories lack. Another difference is that our perception of objects requires continuous movement, while memory is often discontinuous. In other words, while we have to pass through all the space separating us from an absent object in order to perceive it, we can leap into the past, over great spans of time, when we remember. According to Bergson, these differences are the source of the notion that "unperceived objects in space and unconscious memories in time [are] two radically different forms of existence" (MM 146/288). In his view, it would be more accurate to say that "the exigencies of action are the inverse in the one case of what they are in the other" (MM 146/288). This difference notwithstanding, we have no more reason to believe in the existence of absent objects than we do in the existence of unconscious memories.

Bergson's examination of pure memory and the prejudice against believing that the past exists leads him to raise the ontological question of what exactly it means for something to exist, which he calls "the capital problem of *existence*" (MM 146/288).[26] The disagreement between the theories on opposite sides of the mind-body debate reveals two different conditions for existence, according to Bergson. Idealists view existence as either partially or completely dependent on "presentation to consciousness," while realists view existence as determined by "the logical or causal connection of that which is so presented with what precedes and what follows" (MM 147/288). However, Bergson argues, existence implies both: "The reality for us of a psychological state or of a material object consists in the double fact that our consciousness perceives them and that they form part of a series, temporal or spatial, of which the elements determine each other" (MM 147/288). In his view, for something to exist, whatever it is, it has to meet both conditions to some degree. In most cases, these two conditions are neither completely nor equally fulfilled. With regard to a current psychological state, for instance a visual sensation, its presentation to consciousness may be perfect, but its connection with other things, such as the source of light, is imperfect; in other words, the state itself is absolutely evident, but its cause is not. On the contrary, with regard to a nearby material object, its logical or causal connection to other objects is perfect, but its presentation is imperfect. For instance, a swinging pendulum adheres completely to the law of gravity, but its weight is something that we need instruments to measure. Nevertheless, thanks to the intellectual habit of preferring clear-cut distinctions, Bergson argues, we tend to dissociate the two conditions and regard external objects and internal states as displaying "radically different modes of existence" (MM 147–8/289), one defined by conscious apprehension and the other by causal determination.

The dissociation of these two conditions for existence is precisely what leads realists to attempt to reduce mind to matter and idealists to attempt to reduce matter to mind, according to Bergson. However, each attempt suffers from an illusion based on the overemphasis of one condition. The tenet of idealism that objects cannot exist apart from consciousness "ends by falsifying our representation of matter," he claims, and the tenet of realism that psychological states cannot exist apart from a special kind of matter—cerebral matter—"vitiates our conception of mind by casting over the idea of the unconscious an artificial obscurity" (MM 148/289). Together, these two illusions render both our consciousness of matter and the existence of mind doubtful. We can now see that *Matter and Memory* begins, with the theory of pure perception, by confronting the former problem, and turns, with the theory of pure memory, to the latter. To affirm the existence of mind, Bergson questions whether, for something to exist, it is really necessary for it to be present. If existence depends on some degree of conscious apprehension and regular connection, Bergson claims, then "these two conditions assure to each one of the past psychological states a real, though an unconscious, existence" (MM 148/289). For, in the first place, we are more or less conscious of the "psychological anterior" that is our character. As he suggested in *Time and Free Will*, character is not merely a set of personality traits or predispositions that determine our behavior, but the 'whole soul,' which includes the entirety of our past. In the second place, our choices are influenced by, and therefore connected to, *all* our past experience, most of which we do not consciously recall. Bergson concludes that these are good reasons to believe that memories, no less than objects, exist even without being present in the form of images.

These reflections on existence bear directly on the question of the relationship between the past and the present. We will never understand memory, according to Bergson, until we reject the false problem of where memories are stored. This question is posed in terms of space, and it rests on the assumption that memories are contents in need of a container. The spatial metaphor of containment is inadequate, even absurd in this context, for it demands that something inherently non-spatial be given a definite location in space. Moreover, Bergson claims, in a passage that apparently left a deep impression on the young Heidegger, the metaphor "borrows its apparent clearness and universality from the necessity laid upon us of always opening space in front of us and always closing duration behind us" (MM 148–9/290).[27] In other words, the double movement of opening-up and closing-off is at the basis not only of the illusion that the past is dead, that it *is* no longer, but also of the assumption that whatever exists must be located somewhere in space. This may be true of material objects, but it cannot be true of the events we recall from the past. No present object, not even the brain, is capable of containing the past. The relation of containment is peculiar to objects in space, which exist simultaneously. The question of where memories are stored assumes not only that time can be reduced to space, but also that anything that exists must have a present location. Nothing survives, past or present, without duration, for Bergson, but we constantly confuse "duration itself, in the way it flows" with "instantaneous sections that we apply to it" (MM 149/2910). Therefore, understanding the existence of memory requires a shift in focus from where and how memories are stored to how the past survives "in itself."

According to Bergson, while most of our distant past remains in the shadow of forgetting at any given moment, our present is nevertheless so permeated by the past that "every perception is already memory" (MM 150/291). This is the case with actual (as opposed to pure) perception, because all actual perception endures; that is, the objective, impersonal outlines of present things are filled in with detail and given continuity through the recollection of past experiences. Why, then, do we assume that the present and only the present exists? For Bergson, this common view is no less paradoxical than the view that the past continues to exist:

> Nothing *is* less than the present moment, if you understand by that the indivisible limit which divides the past from the future. When we think this present as going to be, it exists not yet, and when we think it as existing, it is already past. If, on the other hand, what you are considering is the concrete present such as it is actually lived by consciousness, we may say that this present consists, in large measure, in the immediate past. (MM 150/291)

Not only are perception and memory mixed, but what is given in an instant is only a miniscule part of actual perception. Memory plays such a fundamental role in perception, for Bergson, that he goes even further: "*Practically, we perceive nothing but the past*, the pure present being the invisible progress of the past gnawing into the future" (MM 150/291). However, our interest lies mainly in present objects, especially the ones on and with which we can act. This interest in what is present, or actual, prevents us from admitting "the integral survival of the past" (MM 150/291).

In the course of this "long digression" on the problem of existence, Bergson attempts to challenge the habitual representation of the past as having ceased to exist. We tend to think of things in the past as no longer existing, he claims, because they are not useful, thereby turning the practical distinction between what can be acted upon and what cannot into a metaphysical one. But, if present perception is always shaped and guided by memory, so that the present is nothing without the past, what sense does it make to believe that only the present exists? To understand pure memory, Bergson contends, we need to break the habit of equating existence with actuality—of being with being present—and to recognize that *to be no longer* is not the same as *not to be*. Instead of saying that the past no longer exists, we should say that it exists virtually, as opposed to actually. As he later explains in "The Perception of Change," this way of thinking is opposed to a long tradition that privileges the present:

> We are inclined to think of our past as inexistent, and philosophers encourage this natural tendency in us. For them and for us the present alone exists by itself: if something of the past does survive it can only be because of help given it by the present, because of some act of charity on the part of the present, in short—to get away from metaphor—by the intervention of a certain particular function called memory, whose role is presumed to be to preserve certain parts of the past, for which exception is made, by storing them in a kind of box. (CM 150–1/1385)

Not only does the traditional way of understanding memory suffer from spatial thinking, assuming that memories must be stored in the brain, but it also suffers from the ontological prejudice that whatever exists must be present in some form. On the basis of this assumption, memories have traditionally been regarded as ideas imprinted on the mind or lying dormant in the brain. Bergson argues, on the contrary, that memory depends on the survival of the past in itself, or its virtual coexistence with all that is present and actual.[28]

6. The Movement of Memory

With his affirmation of the virtual existence of the past, Bergson poses a radical challenge to the ontological prejudice that whatever exists must be present in some form. His discussion of the problem of existence sheds some light on what it means to say that something exists and shows that we have good reasons to rethink the notion of the past as having ceased to exist. On his analysis, although presentation to consciousness is traditionally held to be one of the conditions for existence, it is not fulfilled in the case of unperceived objects. The other condition for existence, logical or causal connection, which supports the view that absent objects exist, is also fulfilled by unconscious memories. It is tempting to ask where such memories exist, but such a question is based on the habit of thinking spatially and, as we now see, equating being with being present. Memory-images are present, in Bergson's view, but pure memory belongs to a past that was never present.[29] So the question is not *where* memories are located, which assumes that they are contents in need of a container, but *how* the past survives in itself. Accordingly, pure memory must be understood as a process or movement by which the past exists even when it is not illuminated by consciousness.

In his account of the movement of memory, Bergson uses another diagram, the "cone of memory," to describe how, as opposed to where, the past survives. He depicts "the totality of the recollections accumulated in my memory" (MM 152/293) as an upside-down cone whose base symbolizes the past, where all the events of our lives are preserved in minute detail, and whose point or summit symbolizes the present, where we find the "sum of the sensori-motor systems organized by habit" (MM 152/293)—that is, the body. At its summit, the upside-down cone touches the "plane of action," moving across it as time passes, and this represents the "connecting link" between the present and the past. On the plane of action, the body is a "place of passage" that "receives and restores actions emanating from all the images of which the plane is composed" (MM 151–2/292–3). This new diagram incorporates Bergson's previous illustration of the intersection between time and space, with the vertical line becoming the cone of memory and the horizontal line becoming the plane of action. In addition to movement on the plane of action, which includes everything extended in space, thus all matter, there is also movement in the upside-down cone, which turns out to represent mind or spirit. Although Bergson has argued that true memory and bodily memory are different in kind, this dynamic relationship between them illustrates how the past survives and permeates the present without being contained in it.

The purpose of Bergson's cone diagram is to assist in thinking about memory in temporal terms, as a movement involving both true memory and bodily memory.[30] On his account, true memory offers the lessons of experience to the motor mechanisms of bodily memory, while those mechanisms give unconscious memories a way to actualize themselves. The cone helps us to picture this exchange as a *contraction* of the past into the point of action—where our bodies adopt attitudes that allow memories not just to appear to consciousness, but also to materialize into actions—and an *expansion* of the present into contemplation—where our minds turn away from utility and attention to life. Bergson's discussion of attention anticipated this process in the exchange he described between an object of perception and the memory-images it calls forth. He now characterizes this movement as an oscillation between two psychological extremes, pure contemplation and pure action: "We tend to scatter ourselves out over [the base of the cone] in the measure that we detach ourselves from our sensory and motor state to live in the life of dreams; we tend to concentrate in [the summit] in the measure that we attach ourselves more firmly to the present reality, responding by motor reactions to sensory stimulation" (MM 162–3/302).[31] Thus, the cone symbolizes not only the relationship between the two forms of memory, but also two extremes of mental life.

Bergson identifies the base of the cone, representing the totality of unconscious memories, with dreams, in which ordinarily useless images appear thanks to a relaxation of the mechanisms by which true memory is inhibited. In dreams, he argues, memory-images return without any practical need because when we are asleep, our attention is no longer fixed by the "sensori-motor equilibrium of the body" (MM 174/).[32] Similarly, "living in the past" to an impractical extent and lingering over memories with no use for present action are marks of a "dreamer." At the opposite extreme, bodily memory rules, leading to the immediate reactions and the lack of deliberation that characterize an impulsive person. Bergson imagines two individuals, one inhabiting the base and the other the summit of the cone:

> A human being who should *dream* his life instead of living it would no doubt thus keep before his eyes at each moment the infinite multitude of the details of his past history. And, conversely, the man who should repudiate his memory with all that it begets would be continually acting his life instead of truly representing it to himself: a conscious automaton, he would follow the lead of useful habits which prolong into an appropriate action the stimulation received. (MM 155/295–6)

Between these extremes lies the "man of action," who recalls the past experiences that best enable him to navigate the present situation, enjoying the condition Bergson describes as "good sense" and "a well-balanced mind" (MM 153/294). In his view, this "happy disposition of memory" depends on an appropriate level of *tension* between the expansion of consciousness into the past and the contraction of past experience into the present.

Bergson's cone of memory thus helps him to describe not only the extremes of mental life, but the movement of thought in general, including the formation of ideas. The ability to think, to speak, and even to hear and understand language depends on memory, but more precisely, for Bergson, it depends on the coordination of action with recollection. The circuit of attention and the double movement of memory (which he also calls a "current") offer a model, in contrast to the associationist theory of the mind, of how present perception is informed by memories of the past.[33] In opposition to the associationist view of ideas as inert, passive, ready-made things, Bergson focuses on the continual process by which our bodies receive and return movements—that is, by which we *act*—and at the same time, "our entire personality, with the totality of our recollections, is present, undivided within our actual perception" (MM 165/305). Our ideas and our actions are the results of this process, and what needs to be understood are not the laws of association that connect ideas to one another, but the process of thought, the "double movement of contraction and expansion by which consciousness narrows or enlarges the development of its content" (MM 166/305). This movement is how the complex motor apparatus of the body is coordinated with similar situations from the past and thus how the past is actualized in the present.

On Bergson's model of the mind, unconscious memories are not actual images, dormant in the brain, but rather "a thousand repetitions of our psychical life" (MM 162/302). Between the limits of mechanical action and disinterested contemplation, that is, there are thousands of levels into which memory can expand. In the diagram of the cone, distance from the point of action symbolizes detachment from the present, but not necessarily remoteness in the past.[34] Our entire past exists at the summit of the cone as well as the base, but the closer we get to the base, the closer our memories come to past events in their lived duration. Through the movement of contraction, memories are pressed into action, losing their individuality and taking a more common and useful form; through the movement of expansion, our attention to life relaxes, and memories take a more personal form. Memories are therefore "an indefinite multitude of possible *states* of memory," and these states are not separate and distinct experiences following one another in a sequence, but "so many repetitions of the whole of our past life" (168/307). These repetitions are "possible reductions" of the past, from the vaguest feeling of resemblance to the sharpest, most detailed images. Between the extremes of the contraction and the expansion of memory, between automatic reaction and endless contemplation, there is always a wide range of possible "degrees of tension of the memory" (MM 169/308), which can also be understood as different "planes of consciousness" (MM 241/371). Rather than being static, the mind is capable of a wide range of mental "tones" or dispositions, depending on the tension between our past and our present.

We can now see how radically Bergson challenges the traditional understanding of the relationship between the past and the present. He argues that memory-images partake of perception and therefore are not "pure." Pure memory must be limited to the past as strictly as pure perception is limited to the present. With the cone of memory, Bergson offers a model for thinking of memory as a movement of expansion and contraction through which the past survives in itself, whether or not we are consciously aware of it.[35] In this way, he interprets memory as not only a faculty that allows the past to return to the present, but as the *existence of the past in itself*. The "repetitions of our psychical life" are more than just reenactments of past events; they are those events themselves, preserved in such a way that they can return in the form of either conscious images (true memory) or unconscious movements (bodily memory). If we must have a spatial metaphor, we should reverse the common view of memory and say that *the past contains the present*, which is its most contracted point. In terms of time, though, we can say simply that *the past repeats itself at every moment*, or that we always are, at every moment, a repetition of what we have been.[36]

7. Rhythms of Duration

Bergson's challenge to the privilege of presence culminates in chapter four of *Matter and Memory*, where he returns to the question that motivated his examination of memory, that of the relationship between mind and body. His ultimate aim with respect to the mind-body problem is to show how the difficulties of "ordinary dualism" can be overcome with a temporal interpretation of matter and mind (or spirit). In his view, understanding these phenomena in terms of time can account for both their independence, preserving true differences in kind, and their unity, revealing how they coincide. He has argued that pure perception, abstracted from memory, is no different in kind from matter, while pure memory, abstracted from perception, is "absolutely independent of matter" (MM 177/315). Perception and memory, therefore, *are* different in kind: "It is in very truth within matter that pure perception places us, and it is really into spirit that we penetrate by means of memory" (MM 180/317). Bergson's task is to "suppress" the oppositions previously introduced in *Time and Free Will*—the extended and the unextended, the quantitative and the qualitative, and the homogeneous and the heterogeneous—without falling back into realism or idealism. The key to a more tenable dualism for Bergson is to continue thinking temporally. In his view, "[t]he mistake of ordinary dualism is that it starts from the spatial point of view," creating an unbridgeable gap between consciousness and the material world, so "the distinction between body and mind must be established in terms not of space but of time" (MM 220/354). Consequently, Bergson has not only drawn attention to the role of bodily action in perception and memory, but he also has offered temporal interpretations of both, distinguishing them on the basis of the difference between the past and the present. Now he must show how the union of mind and body can be understood in terms of time rather than space while maintaining their independence.

For most of *Matter and Memory*, duration has been in the background of the discussion, but now it finally takes center stage. This is because understanding the distinction between mind and body temporally means thinking in terms of duration, for Bergson. As in *Time and Free Will*, he argues that we need to suspend certain habits of thinking, particularly those that display a "bias in the direction of our utility" (MM 184/321). Previously he traced the problem of free will back to the assumption that time can be adequately represented in terms of space and developed a new concept of time, *pure duration*. Now, once again, the method that Bergson recommends is "to replace ourselves in pure duration, of which the flow is continuous and in which we pass insensibly from one state to another" (MM 186/322). But if duration is the flow of consciousness—a qualitative and heterogeneous multiplicity not to be confused with any quantitative or homogeneous multiplicity—and only consciousness endures, would there not still be an unbridgeable gap between mind and matter? How could material things, extended in space, have any contact with an unextended consciousness? Bergson's theories of pure perception and memory constitute a temporal interpretation of *mind*, but he believes that it is necessary to understand *matter* in terms of duration as well.

To show how matter can be understood in terms of duration, Bergson begins with movement. On his account, movement is something we immediately experience as undivided but habitually represent as divisible into parts. Thus, there is a fundamental difference between concrete movement, which occurs in duration, and the path traversed by a moving thing, which occurs in space. Moreover, it is misleading to even speak of duration as having moments, for although "we distinguish moments in the course of duration, like halts in the passage of the moving body" (MM 190/325), in reality duration never comes to a halt. As soon as we divide duration, projecting it into space and symbolizing it as a line with a beginning and an end, we think of it as having moments or instants corresponding to points on a line. This is the habit of representation that generates Zeno's paradoxes and the problem of free will, whereby, for the sake of utility, we replace "living movement" with "a dead and artificial reorganization of movement by the mind" (MM 193/329). Assuming that there is real movement, Bergson thinks, if we can distinguish between it and our abstract representations of it in terms of the difference between duration and space, then we can similarly distinguish between matter and abstract representations of it.

From the standpoint of ordinary dualism, matter is understood in terms of homogeneous space. It is external, essentially extended, and has nothing in common with mind. It is discontinuous and divisible into parts, or bodies, which act reciprocally on one another by necessity. These bodies are passive things that move and change, but only in response to the action of other bodies. How could anything like a mind, or even a living body, exist in a material world of this sort? Bergson contends that if we think in terms of duration, however, matter lies on a continuum with everything we call "mental" or "spiritual" without being reducible to mind. As with movement, understanding matter in terms of duration means thinking of it as continuous rather than divided into parts. Consider, Bergson says, how physicists have come to regard matter in terms of force as more than mere spatial extension, but rather extension pervaded by "*modifications, perturbations*, changes of *tension* or of *energy*" (MM 201/337). This shows that it is possible to think of matter not as passive and inert, but as *vibrating*, moving according to its own rhythm.

Bergson reaffirms the distinction drawn in *Time and Free Will* between duration and time as a homogeneous medium, claiming that "[t]he duration lived by our consciousness is a duration with its own determined rhythm, a duration very different from the time of the physicist" (MM 205/340). He has shown that with the expansion and contraction of memory, conscious experience involves a wide range of degrees of tension, or tones of mental life, between automatic action and disinterested contemplation. As such, he claims, there are different *rhythms* of duration within consciousness, from the automatism of the "man of impulse" to the reverie of the "dreamer." Indeed, there is a vast spectrum of rhythms, making time flow differently depending on whether we are performing a mindless task, attempting to solve a complex problem, playing a sport, hiking through the woods, having a conversation, waiting for an appointment, and so on. There are not just two kinds of time, the concrete duration of consciousness and the abstract time of physics. Rather, there is a single, unified, homogeneous time, while there is an irreducible plurality of rhythms of duration.[37]

Not only are we conscious of different rhythms of duration, Bergson proposes, but beyond human experience, there also must be "different rhythms which, slower or faster, measure the degree of tension or relaxation of different kinds of consciousness and thereby fix their respective places in the scale of being" (MM 207/342). Accordingly, duration is not just a subjective feature of human experience, but also an essential characteristic of beings both "higher" and "lower" than ourselves, which have their own distinctive temporal rhythms. Even matter, insofar as it consists of vibrations, can be considered as having its own rhythm of duration. According to Bergson's earlier analysis, "we seize in the act of perception, something which outruns perception itself, although the material universe is not essentially different or distinct from the representation we have of it" (MM 208/343). Pure perception is thus a consciousness of material objects that adds nothing to them, but subtracts what is uninteresting for us. Going further, actual perception, thanks to memory, also "contracts into a single moment of my duration that which, taken in itself, spreads over an incalculable number of moments" (MM 208/343). In other words, our present consciousness condenses innumerable qualities, innumerable sensations, into its representations. If we abstract from our consciousness of material objects to imagine matter in itself, Bergson argues, we do not necessarily eliminate duration altogether, but only "that particular rhythm of duration which was the condition of my action upon things" (MM 208/343). Even apart from consciousness, matter moves continuously, and the qualities that we would perceive exist in the form of "vibrations, all linked together in uninterrupted continuity, all bound up with each other, and traveling in every direction like shivers through an immense body" (MM 208/343). On this account, matter has its own rhythm of duration, which we overlook when we reduce it, as Bergson had in *Time and Free Will*, to nothing more than extension in space.

The traditional distinction between matter as extended in space and mind as unextended leaves us few options for understanding their interaction. Bergson suggests replacing this spatial distinction with a temporal one and thinking of two kinds of rhythm, one "a succession of infinitely rapid moments which may be deduced each from the other and thereby are *equivalent to one another*," and the other "being in perception already memory, and declaring itself more and more as a prolonging of the past into the present, a *progress*, a true evolution" (MM 221/354). The first would be matter as pure repetition, without indeterminacy: "If matter does not remember the past, it is because it repeats the past unceasingly, because, subject to necessity, it unfolds a series of moments of which each is the equivalent of the preceding moment and may be deduced from it: thus its past is truly given in its present" (MM 223/356). This rhythm of duration is "rapid" in the sense that there are no delays between action and reaction, no intervals for memory to insert images from the past; the tension of memory that we find in ourselves when we act mechanically is present in the highest degree in matter, such that the past is only acted, never represented.

If matter is pure repetition, a more complicated rhythm characterizes mind (or spirit), "distinct from matter in that it is . . . *memory*, that is to say, a synthesis of past and present with a view to the future" (MM 220/354). While matter tends toward mechanical reaction and repetition, on Bergson's account, the mind tends toward creativity and difference. Now freedom can be defined more precisely in terms of the difference between these two kinds of rhythm:

> To reply, to an action received, by an immediate reaction which adopts the rhythm of the first and continues it in the same duration, to be in the present and in a present which is always beginning again—this is the fundamental law of matter: herein consists *necessity*. If there are actions that are really *free*, or at least partly indeterminate, they can only belong to beings able to fix, at long intervals, that becoming to which their own becoming clings, able to solidify it into distinct moments, and so to condense matter and, by assimilating it, to digest it into movements of reaction which will pass through the meshes of natural necessity. . . . The independence of their action upon surrounding matter becomes more and more assured to the degree that they free themselves from the particular rhythm which governs the flow of this matter. (MM 210/345)

As perception contracts matter in the present, memory prolongs the past into the present, opening up the field of possible action. Bergson notes that the capacity for memory varies with the complexity of a living being's body, and especially its nervous system, which "allows the being to free itself from the rhythm of the flow of things and to retain in an ever higher degree the past in order to influence ever more deeply the future" (MM 222/355). Our freedom corresponds in this way to our power of recollection: rational action and intelligent reflection are limited by the past at our disposal. This description of freedom at the end of chapter four of *Matter and Memory* is consistent with the account offered in *Time and Free Will*, with free acts drawing from our entire past, expressing and changing the fundamental self, being rare, and being a matter of degree. Now, in terms of duration, Bergson says, "we can conceive an infinite number of degrees between matter and fully developed spirit—a spirit capable of action which is not only undetermined, but also reasonable and reflective" (MM 221/355), for between matter and spirit "there are all possible intensities of memory or, what comes to the same thing, all the degrees of freedom" (MM 222/355). Understood in terms of duration, mind thus differs from matter by virtue of the tension that memory maintains between the past and present.

In the final analysis, Bergson's new dualism resolves the mind-body problem by rethinking both their independence and their interaction. On the one hand, body and mind differ fundamentally, the body being "an instrument of action and action only" (MM 225/356) and the mind being "a movement between action and representation" (MM 243/372), namely the oscillation of memory. But they are by no means two "separate and distinct" substances; understood in terms of time rather than space, the body is situated in the present, on the plane of action, and the mind moves back and forth between the past and present with a view to the future. On the other hand, body and mind continually interact, the body inhibiting memory and thereby limiting the mind to useful representations, and the mind recalling past experiences that enrich perception and inform action. Thinking in terms of duration, Bergson revisits several of the oppositions he had affirmed in *Time and Free Will* and shows how they can be reconciled. First, his theory of matter reveals that the opposition between the extended and the unextended is a result of spatial thinking. Prior to the distinction between objects in homogeneous space and sensations in consciousness, thus prior to extension in the traditional sense, or "extensity" (*étendue*), there is the extension (*extension*) of pure perception. If, as Bergson has argued, perception and matter coincide in pure perception, then matter "is not, cannot any longer be, that composite extensity which is considered in geometry; it indeed resembles rather the undivided extension of our own representation" (MM 182/318). The matter with which perception coincides is not pure extendedness in space, but continuous movement, vibration, a fixed rhythm of duration.

Bergson claims that the opposition between quality and quantity, as well as between heterogeneity and homogeneity, becomes less pronounced when not only sensations but also objective changes are understood in terms of duration. Insofar as memory condenses an "incalculable number" of moments of the duration of matter into one moment of our duration, "the heterogeneity of sensible qualities is due to their being contracted in our memory and the relative homogeneity of objective changes to the slackness of their natural tension" (MM 182–3/319). Prior to our conception of it as traversing homogeneous space, the vibration of matter, like the qualitative changes of consciousness, prolongs the past into the present and therefore "already possesses something akin to consciousness" (MM 247/376). Between sensible qualities and calculable changes, there is "only a difference of rhythm of duration, a difference of internal tension" (MM 247/376). Ultimately, the problem of how a moving body can produce sensations in the mind of one who perceives it can be traced back to the substitution of abstract motion in space for concrete movement in duration. Spatial thinking reduces the different degrees of tension and extension to the dichotomy between the extended and the unextended.

Finally, as we have just seen, the opposition between freedom and necessity can be overcome by understanding matter in terms of duration. For Bergson, this means thinking of the mathematical deducibility of one moment from the one preceding it as the "rhythm of necessity" and recognizing that the capacity for choice is a result of evolution, specifically the development of living beings with complex central nervous systems that modulate that rhythm. What is visible, with regard to natural evolution, is the development of beings with "an ever greater latitude left to movement in space," and what is invisible, but no less essential to freedom, is "the growing and accompanying tension of consciousness in time" (MM 248/377). This tension between action and recollection is what enables complex organisms to "trump" necessity, so to speak. However, Bergson argues, "whether we consider it in time or space, freedom always seems to have its roots deep in necessity and to be intimately organized with it" (MM 249/378). The opposition between freedom and necessity, like the others, can be overcome not only by thinking in terms of duration, but also by following duration beyond human consciousness, to rhythms of duration other than our own.

8. Ecstatic Duration

If we pay attention to the development of Bergson's thinking in *Matter and Memory*, we can see that Heidegger had to ignore this work in order to characterize duration as a feature of our consciousness of time. While there are some obvious similarities between Bergson's philosophy of time and Husserl's phenomenology of internal time-consciousness, his interpretations of consciousness and duration undergo not just an evolution, but also a crucial transformation. In *Time and Free Will*, Bergson says nothing to discourage the traditional view of consciousness as an inner realm of representations. His distinction between lived time (pure duration) and time as it is represented appears to mirror Husserl's distinction between subjective, immanent time and objective, transcendent time. Bergson also emphasizes that the essence of time, as opposed to space, is to pass unceasingly, or to flow. However, in *Matter and Memory* he systematically calls into question the being of consciousness and the subjectivity of duration.

In the course of rethinking the relationship between mind and body, Bergson makes a number of unorthodox claims about consciousness: it is not primarily speculative, but practical; it is neither subjective nor internal without the contributions of memory; it is conditioned by the unconscious survival of the past; it consists of many levels or planes between pure action and pure contemplation; and it is prefigured in the vibrations of matter. Through his study of memory, Bergson completely rethinks the notion that duration belongs only to our conscious experience. Just as within consciousness there are various degrees of tension between the past and the present, he claims, there are various rhythms of duration beyond consciousness. As such, the duration of human consciousness, our experience of the flow of time, is open to other durations.[38] As something that characterizes the rhythmic organization of mind as well as matter, duration is not just psychological, but also ontological. Rather than duration being a property of consciousness, consciousness is a product of duration.

It is not hard to see what Heidegger must have considered valuable in *Matter and Memory*: Bergson's path leads to a direct confrontation with the privilege of presence. Thinking temporally, Bergson characterizes perception and matter as different in degree only, since both occupy the present, but perception and memory as different in kind, since memory preserves the past. Bergson resists both realism and idealism, much as Heidegger does in *Being and Time*, by rejecting the epistemological prejudice that perception is directed primarily toward knowledge. For Bergson, the difference between being and being consciously perceived must be understood practically: conscious perception is virtual action. When he turns from perception to memory, Bergson confronts the ontological prejudice that only what is present can exist. He argues that the basis of the metaphysical distinction between things that presently exist and things in the past that are thought no longer to exist is the practical difference between things we can act on and things we cannot. Instead of saying that what is past no longer exists, he says that it exists virtually. The past is not just what has passed by and therefore *is* no longer, but what survives and repeats itself, maintaining a dynamic tension with the present. Not only is everything present to consciousness permeated by the past, but without this tension, there would be no presence.[39]

Although Heidegger criticizes Bergson for neglecting the question of being, Bergson makes a definite turn toward ontology in *Matter and Memory*. He first raises the question of the being of consciousness by asking about the difference between being and being consciously perceived. Although consciousness is indeed his thematic field in *Time and Free Will*, where he defines duration in terms of our lived experience of time, in *Matter and Memory* he problematizes his earlier stance with an account of how memory informs perception. Not only does he reject the notion of consciousness as subjective and internal when he thinks about pure perception, but when he thinks about pure memory, Bergson also rejects the equation that being equals presence. He challenges the privilege of presence by arguing in favor of the existence of unconscious memories, which are not stored in the brain and are never actually present. In this way, Bergson recognizes something not sufficiently considered: that being in the past is altogether different from not being at all. "To be no longer" means not to be active, which is not the same as not to exist.

Bergson's ontological turn in *Matter and Memory* complicates and deepens his understanding of time. When Bergson speaks of pure memory, he is not thinking of a mental faculty; that is, he is not thinking of it in psychological terms. Rather, in ontological terms, *pure memory is the mode of being of what has been*.[40] Even though most of what has been is usually inhibited from returning, or forgotten, the past is represented in images (true memory) and repeated in actions (bodily memory). This anticipates Heidegger's interpretation of the "having-been" (*Gewesenheit*) of Dasein, whose modes are forgetting (*Vergessenheit*) and repetition (*Wiederholung*) (BT 324/339). For Bergson, the past is preserved as a whole, never destroyed, and even what is forgotten exists virtually, pressing into the present and awaiting opportunities to return. Moreover, actual perception is always mixed with recollection, and the contributions of memory are so great that Bergson thinks it is no exaggeration to say that most of what we perceive is from the past. He therefore rejects the notion of the present as a mathematical instant as a mere abstraction, defining it instead as a system of sensations and movements belonging to a lived body. The present is open to the future, as perception prepares for action, but also to the past, as memory informs perception and deepens attention. In this respect, what we call "now" is Janus-faced, for Bergson, providing a passageway or point of contact between action and recollection. Unfortunately, Heidegger does not acknowledge the ways that Bergson's theory of pure memory poses a direct challenge to traditional ontology and to the concept of "now-time."

If we accept Heidegger's suggestion that Bergson attempts but fails to shed light on originary temporality, then we need to consider not only Bergson's concept of duration, but also his theory of memory. In particular, we need to recognize that Bergson does not rest with his initial descriptions of duration as a continuous, heterogeneous multiplicity of conscious states. For the source of this continuity and this heterogeneity is memory, including both the bodily memory that prolongs the present through action and the true memory that completes perception with recollection. While Bergson does describe duration in *Time and Free Will* as a succession of conscious states, in *Matter and Memory* he goes beyond this level of description to focus on how perception is enriched and deepened by memory. He describes the movement of memory as a current traveling along a circuit between objects and recollections, an oscillation between the past and the present. In this back-and-forth movement, we find something closely resembling what Heidegger calls the "temporalization" (*Zeitigung*) of time.[41] Bergson discovers this movement by thinking about the mind in terms of duration rather than space, and he comes to think of conscious experience as involving different tones of mental life or degrees of tension. He concludes that there are many different rhythms of duration, both within and beyond human experience.

Thus, when we take *Matter and Memory* into account, we can see that Bergson thinks of duration not simply as a feature of consciousness or interiority, but as a movement toward the outside, an *ecstasis* in Heidegger's sense. Thanks to memory, not only are we conscious of our past, a past that presses into the present, but we are also capable of many degrees of tension between the past and the present, and many rhythms of duration. For Bergson, our duration is ecstatic, open to durations other than our own, so we are "outside ourselves" in a world inhabited by not just material objects, but also different rhythms of life. There is no longer a dualism between consciousness and the external world because our duration encompasses the duration of things, and things endure both "above" and "below" us according to their rhythm. The present is nothing without the past, which repeats itself at every moment. With his theory of memory, Bergson appears to have done much more than merely reverse either the traditional concept of time or our ordinary way of understanding it. In fact, he has introduced some of the very features of originary temporality that Heidegger presents as his own discoveries in *Being and Time*.

Conclusion

The Movement of Temporalization

1. The Swinging of Time

According to Heidegger's analysis in *Being and Time*, division two, the ecstatic unity of temporality is concealed by our tendency to think of time as something objectively present, namely a continuous, irreversible succession of nows. This is the ordinary understanding of time that guides philosophers from Aristotle to Bergson. Ever since Aristotle emphasized the connection between time and the mind and treated the "now" as the only part of time that exists, philosophers have vacillated between thinking of time as an objective feature of the world and as a subjective feature of human experience. In Bergson's attempt to radically distinguish time from anything objective or extended in space, Heidegger claims, he, no less than Husserl, thinks of duration as subjective.[1] As Heidegger says in *The Metaphysical Foundations of Logic*:

> Recently Bergson tried to conceive the concept of time more originally. He made it more clear than any previous philosopher that time is interwoven with consciousness. But the essential thing remained unresolved in Bergson, without even becoming a problem. He developed his interpretation of time on the basis of the traditional concept of consciousness, of Descartes' *res cogitans*. The basic metaphysical problem of the primordial connection between Dasein and temporality he does not pose, and even less does he pose the problem of being, for which the other problem is only a preparation. (MFL 149/189)

Thus, in Heidegger's view, no matter how successfully Bergson distinguishes duration from anything spatial, he fails to clarify temporality as the movement of transcendence by which Dasein is "outside of itself" in the world.

Beginning with the immediate data of consciousness, Bergson's analysis is limited to the human experience of time, and therefore he follows Kant, Augustine, and Aristotle in recognizing the connection between time and the subject without posing the question of "the subjectivity of the subject."

Ultimately, for Heidegger, temporality cannot be understood properly as "a psychical process or a property of a psychical process" (MFL 202/260). Any such process or property would be something objectively present, even if it were thought to be "in the subject." Instead, temporality needs to be understood in terms of the movement expressed in the utterances "now," "then," and "formerly." This movement is not simply the flow or stream of time, which is merely a common representation of now-time. Nevertheless, in an aside that displays Heidegger's characteristic ambivalence toward Bergson, he writes:

> N.B. Bergson first worked out the connection between a derived and an original time. But he did so in a way that went too far and said that time, once emerged, is space. Bergson thereby blocked the way to the real understanding of derived time, since he, in principle, mistakes the essence of emergent time, insofar as he does not view as emergent the time that has emerged. But, conversely, insofar as he stays merely within the time that has emerged, he does not really succeed either in clarifying primordial and genuine time in its essence. Bergson's analyses nonetheless belong to the most intense analyses of time that we possess. (MFL 203/262)

The "derived" time to which Heidegger refers is the sequence of nows, which "emerges," on his account, from the intentional unity of awaiting, retaining, and making-present. What Heidegger calls "original time" is, of course, ecstatic temporality, which shows up in the everyday mode of "the making-present that awaits and retains." As he had argued in *Being and Time*, Bergson's interpretation of original time as duration covers up both the ecstatic-horizonal structure of temporality and the process of leveling-down by which it becomes now-time. Here, he suggests that Bergson's main contribution was to have recognized in a general way the difference between time and temporality, but his failure was to have misunderstood both by treating time as an illegitimate, spatial representation of duration.

Heidegger goes on in *The Metaphysical Foundations of Logic* to credit Husserl with discovering the unity of awaiting, retaining, and making-present, but first he cautions against dismissing Bergson too hastily:

> It has become a commonplace that Bergson (as well as Dilthey) is fuzzy and must be therefore re-examined and improved. But Bergson's "images" are the very expression of his exertions to really grasp the phenomenon within the realm he takes for his theme. The lack does not lie in an alleged fuzziness—Bergson is perfectly clear in what he sees. But it lies in the overly narrowed realm of his set of problems. Nor would this be removed by a revision for "greater exactness." As everyone knows, there are also exact trivialities in philosophy! (203/262–3)

With this, Heidegger denies that Bergson's thinking is simply "fuzzy," but also that it could be salvaged through attention to Bergson's call for greater precision in philosophy.[2] He suggests that Bergson grasps something with clarity and precision, but not the *primal something* from which our ordinary understanding of time emerges.

To demonstrate the difference between ecstatic temporality and Bergsonian duration, Heidegger considers one of the ways that Bergson visualizes the relationship between the past, present, and future in pure duration. At the origin of what we call "the past," Heidegger describes an ecstatic movement:

> having-been-ness [*Gewesenheit*] temporalizes itself only from out of and in the future. The having-been [*Gewesensein*] is not a remnant of myself that has stayed behind and has been left behind by itself. Neither is it what Bergson likes to illustrate with various images: the future unrolls, as it were, while the past is rolled up on another roll.[3] . . . the has-been is not something remaining by itself, nor is it an accumulating dead weight I haul behind me and to which I occasionally relate in one way or another. Rather, my having-been only "is," in each case, according to the mode of temporalization [*Zeitigung*] of the future. . . . What we find here regarding the essence of temporality is that the future ecstasis, as a coming-towards, stretches out immediately, constantly, and primarily into the having-been. (MFL 206–7/266–7)

Bergson's image illustrates the continuity, mobility, and indivisibility of duration, but, according to Heidegger, it also reinforces the notion of the past as something bygone, no longer in effect, and behind us. Heidegger could have said likewise that this image represents the future in a misleading way as something ahead of us, moving in our direction. In both cases, the image of the spools represents time as a distinctive movement, but not the movement of ecstatic temporality. In place of this image, Heidegger suggests a question mark with one arrow pointing toward it and another pointing away. This illustrates that Dasein, a being who is always in question, comes toward itself out of its own possibilities (the future) and comes back to its having-been (the past) "out of and in the future." Dasein thus "temporalizes" as a thrown projection, a being that *is* both what it *can be* and what it *has been*.

But here, where the difference between Bergsonian duration and Heideggerian temporality seems clearest, Heidegger goes on to complicate the picture. He acknowledges how difficult it is, thanks to the privilege of presence, to grasp the movement of ecstatic temporality without attributing it to a subject that undergoes that movement. It is difficult to break the habit of asking *what* it is that moves in this distinctive way. Yet it is crucial, Heidegger contends, to avoid the mistaken assumption "that these three ecstasies flow together somehow in one substance . . . something present on hand unecstatically . . . the common center for initiating and unfolding the ecstasies" (MFL 207/268). We will fail to grasp the movement of temporalizing as long as we think of time in terms of *presence*. Although Heidegger repeatedly denies that Bergson succeeds in getting to the root of time, he calls on Bergson for help in reversing our habits of representation: "if we may speak at all about the 'being' of the ecstasies, we must say that it lies directly in the free ecstatic momentum [*Schwung*]. Bergson speaks of this phenomenon with his term 'élan'" (MFL 207/268). Predictably, Heidegger is quick to add that Bergson fails to see the ontological significance of this term and thinks of it as a unidirectional movement, a forward impetus. However, he continues, "Temporalization is the free oscillation [*Schwingung*] of the whole of primordial temporality; time reaches [*erschwingt*] and contracts [*verschwingt*] itself. . . . the essence of time lies in the ecstatic unitary oscillation" (MFL 208/268–9). Heidegger thus characterizes the movement of originary temporality as *oscillating* rather than unrolling, and concludes, "*World entry* happens only insofar as something like ecstatic oscillation temporalizes itself as a particular temporality" (MFL 209/269). It is the swinging back and forth of ecstatic temporality that makes Dasein open to the world rather than closed in on itself, so we can say: "Time is essentially a self-opening and expanding [*Ent-spannen*] into a world" (MFL 210/271).

With this discussion of the "swinging" of temporality, Heidegger demonstrates why it is impossible to see just how radical Bergson's thinking about time is without considering his theory of memory. In *Time and Free Will*, Bergson indeed defines duration as a feature of consciousness, and there is little to suggest that he departs in any significant way from Descartes's dichotomy between *res cogitans* and *res extensa*. Bergson develops his concept of duration by abstracting the flow of time from all extension in space, and in this respect he appears to reaffirm Cartesian dualism: on the side of consciousness, there is succession without externality, and on the side of the world, there is externality without succession.[4] However, in *Matter and Memory* he overcomes this dualism by demonstrating how a conscious being is already outside of itself in perception and how the material world endures thanks to memory. On his account, memory is not just a psychological faculty, but also a back-and-forth movement between the past, which survives on its own but repeats itself in us, and the present. In actual perception, memory-images project themselves at objects, forming a circuit between recollection and action. Bergson's cone of memory, an image that Heidegger overlooks, represents the double movement of expansion and contraction by which the past is translated into movements or images. The past is only "dead weight" from a practical standpoint because it cannot be acted upon. In fact, Bergson maintains that the past survives and animates the present, but he claims that we overlook this insofar as we elevate a practical distinction to a metaphysical one. In short, if the question is whether or not Bergson succeeds in clarifying the movement of temporality, the answer lies not in his concept of duration alone, but in the theory of pure memory with which he deepens it.

2. Rethinking Time

My aim in examining Heidegger's critique of Bergson and Bergson's strategy of resolving philosophical problems by thinking in terms of time has been to illuminate some of the ways in which Bergson is a much more radical thinker than Heidegger allows. In my view, Heidegger's dismissal of Bergson's thinking about time as too traditional ultimately fails not only because there is no good evidence that Bergson begins with Aristotle's concept of time and reverses it, but also for the reason that Bergson anticipates Heidegger's interpretation of Dasein as temporality in a number of crucial ways that can now be spelled out more clearly. Bergson's efforts to overcome the traditional concept of time resemble Heidegger's own in more and deeper ways than Heidegger cared to admit, well beyond the distinction between "derived time" and "original time."

To begin with this distinction, though, the most obvious way that Bergson's thinking foreshadows Heidegger's is through the difference that he establishes between time and duration. On Heidegger's own interpretation, Bergson seeks a more original concept of time, one that goes beyond the traditional concept of time and the ordinary understanding from which it is derived. In this pursuit, Bergson criticizes not just this or that theory of time, but rather certain presuppositions about time that have dominated the history of philosophy. For Bergson, the habit of representing time in terms of space, a tendency rooted deeply in human language and thought, is responsible for some of the basic problems of philosophy. In *Time and Free Will*, he argues that space is implicit in the concept of time as a measurable quantity, which is an abstract, symbolic representation of the lived experience of time, or duration. On Bergson's account, the duration of conscious life is the origin of time as we usually understand it. On the one hand, duration is something that we immediately experience, but on the other, it is covered up by spatial thinking, so "we find it extraordinarily difficult to think of duration in its original purity" (TFW 106/71). In Heidegger's earliest works on time, he employs Bergson's distinction between the homogenous, measurable time associated with physics and a more primordial time that he calls "temporality" rather than "duration." He dismisses Bergsonian duration in *Being and Time*, arguing that it is a mere reversal of Aristotle's concept of time, and at best a feature of time-consciousness. However, in his insistence that Bergson fails to clarify "primordial and genuine time," Heidegger implies that *temporality* is precisely what Bergson sought to reveal with his concept of duration.

Less obviously, Bergson's distinction between homogeneous time and pure duration also anticipates Heidegger's thinking about authenticity. Heidegger criticizes Bergson for operating within an "overly narrowed realm" of philosophical problems, which is another way of saying that Bergson overlooks the question of being. What Heidegger fails to mention is that Bergson attempts to resolve the problem of free will by attacking presuppositions about time and the self shared by both sides of the debate. In particular, he questions the notion that the self consists of a succession of conscious states and the underlying assumption that time can be adequately represented in terms of space. For Bergson, freedom needs to be understood in terms of the living, evolving, and enduring self, as opposed to a self that has been projected into space, carved up into discrete states, and made into "public property." As I have shown, Heidegger's interpretation of authenticity in *Being and Time*, like Bergson's account of freedom, hinges on a distinction between two aspects of the self. For Heidegger, authenticity is a matter of bringing oneself back from the "they-self" to resoluteness or "self-constancy," which he interprets as mode of temporality. Of course, he disagrees with Bergson's diagnosis that the root of the problem is a confusion of duration with space. However, like Bergson, Heidegger attempts to rethink the meaning of selfhood in terms of time and challenges our assumptions about the permanence of the self.

Bergson's account of duration in *Time and Free Will* is to some extent a response to Kant's theory of space and time as the pure forms of sensible intuition. While Bergson wants to establish a fundamental difference between time, or duration, and space, he follows Kant in regarding temporal succession as the result of a mental synthesis whereby sensations are retained and combined with others to produce an experience of continuity such as we have when listening to a melody. On this account, duration is not an objective feature of the world or material things, either in an absolute or a relative sense, but a subjective feature of consciousness. In this respect, Bergson also appears to be in agreement with Husserl, whom Heidegger criticizes for distinguishing between immanent (subjective) and transcendent (objective) time without providing ontological clarification of this distinction. Yet in *Matter and Memory* Bergson's thinking develops in such a way that he ceases to view either consciousness or duration as purely subjective. As he explains, conscious perception is "in things rather than in the mind" (MM 218/352), and material things possess "a real duration and a real extensity" (MM 212/346). Ultimately, for Bergson, mind and matter are distinguished by their different rhythms of duration, so duration cannot be defined as either subjective time or objective time. Instead, what Heidegger says about world time fits Bergsonian duration as well: " 'Time' is present neither in the 'subject' nor in the 'object,' neither 'inside' nor 'outside,' and it 'is' '*prior*' to every subjectivity and objectivity because it presents the condition of the very possibility of this 'prior' " (BT 399/419). The origin of time has to be conceived as prior to the distinction between subject and object.

One of the deepest yet most often overlooked resemblances between Heidegger's thinking and Bergson's is that both challenge, albeit in different ways, the long tradition of regarding reality as timeless or eternal. In *Time and Free Will*, Bergson identifies this way of thinking not only with Zeno, for whom time and change are illusory, but also with Kant, who "preferred to put freedom outside of time and to raise an impassable barrier between the world of phenomena . . . and the world of things in themselves" (TFW 235/153). In later works, Bergson holds up Plato's theory of ideas as indicative of the tendency to devalue time and eliminate duration, a tendency that continues to dominate modern metaphysics and science.[5] In his view, the assumptions that reality is timeless and truth is eternal are based on our preference for that which is stable and unchanging, which keeps Platonism alive and well. Thus, Heidegger's resolution to "understand time in terms of time," his question about the ontological function of time, and his interpretation of "within time" and "outside time" as modes of temporality all have a precedent in Bergson's reading of the history of philosophy. In fact, Bergson challenges not only the traditional concept of time, but the subordination of time to eternity in Western philosophy.

It thus appears as if Bergson and Heidegger share the conviction that radically rethinking time is the key—or at least one key—to transforming philosophy. Although they are obviously not alone in calling for the reformation, revolution, or abolition of philosophy, they distinguish themselves through their attempts to transform philosophical questioning into an activity concerned with the temporality of human existence. Bergson does this in his early works with respect to particular problems: in *Time and Free Will* he shows how the problem of freedom can be resolved by thinking in terms of duration, and in *Matter and Memory* he attempts to reframe the mind-body problem by defining both their independence and their interaction in temporal terms. While these problems are typical of modern metaphysics—problems that revolve around dichotomies such as subject and object, consciousness and world, self and others—Bergson regards them as rooted in the same habits of thinking as the problems of ancient philosophy. He therefore proposes that philosophy undertake a "reversal of the habitual work of intelligence" (CM 177/1410), including a revaluation of reality as moving, changing, and perpetually becoming. Rather than carving things up with ready-made concepts and attempting to reconstitute a fluid reality out of fixed positions, Bergson proposes: "*To philosophize means to reverse the normal direction of the workings of thought*" (CM 190/1422). This requires abandoning or at least questioning many of our assumptions about the practice of philosophy.

Heidegger's efforts to transform philosophy have rarely been compared with Bergson's, even though he follows a similar strategy, at least with respect to temporality. In *Being and Time*, one of Heidegger's aims is to trace the forgetting of being from contemporary thought back through the history of Western philosophy. He views this tradition as united by a concept of time, but more importantly, by an understanding of being that privileges one mode of temporality, *constant presence*. When he interprets Dasein as care, and care as grounded in temporality, Heidegger tries to show how the most persistent and problematic dichotomies of philosophy—again, subject and object, consciousness and world, self and others—can be rethought in terms of time. He deals more directly with the question of being than with

Bergson's "insoluble problems" of metaphysics, but, like Bergson, he seeks a reorientation of philosophy with regard to time. He not only wants to show how ontology is fundamental, but he also aims to uncover the temporal conditions for the possibility of the kind of being that does philosophy: not an isolated, worldless subject, but a thrown projection that temporalizes. His interpretation of temporality is designed to uncover the origin not only of what we call "time," but also of the factical, historical being that we call "human." Like Bergson, Heidegger aims to transform the practice of philosophy by revealing the distinctive temporal structure of human existence and challenging our assumptions about the timelessness of being.

Despite the fact that Bergson does not begin, as Heidegger does, with the conviction that the question of being is *the* fundamental question of philosophy, he does not shy away from ontology. Heidegger's decision to focus exclusively on *Time and Free Will* when criticizing Bergson suits his purposes, because in that work Bergson does not question the Cartesian dichotomy between the inner realm of consciousness and the external world. Yet in *Matter and Memory* Bergson clearly calls into question not only the epistemological prejudice that perception is directed primarily toward knowing, but also the ontological prejudice that being equals presence. First, he attacks the notion of consciousness as an enclosed sphere of representations, and then he rejects the assumption that whatever exists must be present or actual. In fact, Bergson does exactly what Heidegger faults him for failing to do: he questions the being of consciousness and challenges the privilege of presence. On his account, the past survives and animates the present, and the *double movement* of memory makes the presentation of images to consciousness possible. Bergson describes this movement in various ways: as the expansion and contraction of recollection, as the condensation of vibrations and the prolongation of the present, and as the opening up of space and the closing off of the past. Heidegger's struggle, and perhaps his deepest kinship with Bergson, is to reveal this movement as the condition for the possibility of presence, a movement ontologically prior to substance or the subject.[6]

In light of these similarities, we can more clearly see certain genuine differences between Bergson and Heidegger that also deserve our attention. We can begin with an issue that Heidegger seizes upon, namely Bergson's emphasis on the way that space, or spatial representation, is implicit in the concept of time as a measurable quantity. The main problem with the traditional concept of time is not that we think of it in terms of space, Heidegger argues: "Rather, what is ontologically decisive lies in the specific *making present* that makes measurement possible" (BT 397/418). I have shown how Heidegger works to disentangle temporality from spatiality in *Being and Time* and to de-spatialize Aristotle's account of time in *The Basic Problems of Phenomenology*. He thus appears to consider it necessary but not sufficient to distinguish temporality from anything like extension in space. However, on Heidegger's analysis, the concept of pure, geometrical space, which Bergson takes for granted, is derived from a more originary spatiality. He sets the stage for his interpretation of temporality with an existential analysis of spatiality, in which he traces the "pure manifold" of space back to the ontological constitution of Dasein. Although Heidegger does not directly criticize Bergson's conception of space, he could have contested the way that Bergson accepts Kant's view of space without distinguishing it from the concrete spatiality of being-in-the-world. As some critics have pointed out, while Bergson carefully distinguishes between abstract and concrete time, he fails to do the same for space and place.[7] Bergson's account of perception and the lived body in *Matter and Memory* complicates his view of space as essentially homogeneous and abstract, but even in his later works he seems to put all heterogeneity and concreteness on the side of duration. Therefore, it is an open question whether Bergson accepts a traditional concept of space that is as problematic as the traditional concept of time that he opposes.[8]

When we consider the disciplinary and methodological commitments involved in Heidegger's and Bergson's attempts to rethink time, the differences between them multiply. Early on, Heidegger highlights some of his most general concerns by emphasizing Bergson's reliance on psychology and biology. This concern is evident as early as 1919–20, when he associates Bergson and James for their "biological orientation" (PIE 10/15) and criticizes Bergson for overlooking the intrinsic meaningfulness of life "as a result of working from biological concepts" (BPPa 174–5/231). Closer to *Being and Time*, Heidegger suggests that Bergson's "theory of the immediate data of consciousness," a reference to *Time and Free Will*, "goes back to the ideas of Brentano's psychology" (HCT 23/28), and he calls *Matter and Memory* "a basic text for modern biology" (LQT 208/251). Bergson does indeed engage with the empirical sciences in a way that Heidegger refuses to do. In the introduction to *Matter and Memory*, Bergson suggests that psychology and metaphysics be viewed as independent sciences, but that each "should set problems to the other and . . . help it to solve them" (MM 15/167). In contrast, Heidegger insists in *Being and Time* on the distinctness and priority of philosophy's task, for "ontological inquiry is more original than the ontic inquiry of the positive sciences" (BT 10/10). If philosophy is a science—a view that Heidegger eventually rejects—he believes that it must be concerned with *being* rather than *beings*, and it must not take for granted the basic concepts or shared presuppositions of the empirical sciences. Thus, while Heidegger attempts to police the border between philosophy and science, Bergson both draws upon the research of the scientists of his day and challenges some of their deepest assumptions.[9]

Although both thinkers were deeply invested in working out an appropriate method for philosophy, Heidegger's phenomenological method bears little resemblance to Bergson's conception of philosophical intuition. Beginning, once again, as early as 1919–20, Heidegger pursues the problem of method with Bergson as one point of reference along with Natorp, Jaspers, Dilthey, and, of course, Husserl. One of Heidegger's main concerns is to determine how phenomenology can enable us to grasp lived experience without objectifying and thereby "de-vivifying" it. On the one hand, Heidegger agrees that philosophy and science tend to distort lived experience by reifying it and thereby "stilling the stream" of consciousness (TDP 78/101; S 84/19). On the other hand, he denies that the solution depends on a return to the "immediately given," if that means sensations or sense-data. Heidegger vacillates between acknowledgement of Bergson's "partly ingenious intuition" (BPPa 8/10) and insistence that phenomenology recognizes "no special capacity, no exceptional way of transposing oneself into otherwise closed domains and depths of the world" (HCT 47/64). Heidegger famously interprets intuition in *Being and Time* as *noein*, a privileged mode of access to the truth for the ancient Greeks, "the simple apprehension of something objectively present in its sheer objective presence" (BT 24/25–26). For him, the "pure beholding" of intuition is a paradigm of theoretical activity, the "ideal of knowledge" for the entire Western philosophical tradition (BPPb 118/167). In terms of time, he explains, intuition is a kind of *making present* (BT 346n/364n) and therefore cannot provide access to originary temporality.

Although it is safe to assume that Heidegger's critique of intuition was targeted primarily at Husserl, there is much in Bergson's account of intuition that he would reject. In one of Bergson's most famous discussions of method, he calls intuition "a direct vision of the mind by the mind" and "a knowledge which is contact and even coincidence" (CM 32/1273). In contrast to analysis, which seeks knowledge from an external standpoint and "reduces the object to elements already known," intuition, for Bergson, is "the *sympathy* by which one is transported into the interior of an object in order to coincide with what there is unique and consequently inexpressible in it" (CM 161/1395). While the intellect employs concepts, intuition seeks to grasp things "over and above all expression, translation, or symbolical representation" (CM 162/1396). Why would we need to seek such an unmediated view of things? For Bergson, intuition is necessary because the intellect "cannot, without reversing its natural direction and twisting about on itself, think true continuity, real mobility, reciprocal penetration—in a word, that creative evolution which is life" (CE 162/632). If Bergson's view is that life is an irrational and inexpressible flux, which cannot be conceptually understood and is accessible only through intuition, then this marks a clear difference from Heidegger, who thinks of life as *historical* and approaches it hermeneutically.[10]

Heidegger's account of duration clearly differs from Bergson's, but it takes some effort to see through his oversimplifications to why he denies that duration is originary. In *Being and Time*, Heidegger briefly discusses duration with respect to "spannedness," one of the four essential characteristics of world time. On his account, the term "during" expresses a span or interval of time between "now" and "then," which is possible thanks to Dasein being "ecstatically *stretched along*" (BT 390/409). Duration, in this sense, is one way that time shows up in the self-interpretation of Dasein in the mode of the making-present that awaits and retains. In other words, duration is a phenomenon of the time taken care of, the inauthentic temporality of everydayness, as opposed to the authentic temporality of anticipatory resoluteness. Duration cannot account for the ecstatic stretching of Dasein or explain what makes the concept of time as a continuous, irreversible succession of nows possible. Heidegger could even argue, although he does not, that duration as Bergson understands it is still a "succession of experiences 'in time' " (BT 356/373). Having distinguished duration from a quantitative multiplicity like number, from a homogeneous medium like space, and from anything whose parts are separate and distinct, Bergson nevertheless conceives duration as an unceasing flux, the flow of time.[11] Originary temporality does not flow, for Heidegger; it is not the passage of time, but the movement of "temporalization." Temporality is not a blending together of the past, present, and future, but rather a *coming toward* oneself out of the future by *coming back* to one's having-been, the movement of *presencing* as such (BT 310–11/325–6). Duration is not originary, in Heidegger's view, because it is a phenomenon that shows up as we take care of things, not the ecstatic movement that makes care possible.

I have argued that Bergson, even before Heidegger, recognizes this movement, but that we have to look to *Matter and Memory* for his descriptions of it. That said, there are other genuine differences between Heidegger and Bergson on the issue of time that may lead to more fruitful debate. Although I have suggested that with *Matter and Memory* in view, we can see that Bergsonian duration is ecstatic, it is true that Bergson does not regard death as crucial to temporality. As Heidegger is obviously aware, Bergson emphasizes the openness and indeterminacy of the future, and therefore the novelty and creativity of life.[12] Bergson is more interested in the survival of our past, and the possibility that the mind avoids annihilation, than our mortality and our flight from death. Accordingly, in contrast to Heidegger, Bergson offers no account of the uncanniness of being and dismisses the "nought" (*néant*) as a pseudo-problem. That is, while Heidegger maintains that anxiety in the face of death reveals the "nothing" (*Nichts*) at the basis of human existence (BT 180–2/186–8; 266–7/276–7), Bergson argues that the idea of nothing is merely the negation of everything, and the metaphysical question "why is there something rather than nothing?" is a false problem.[13]

These differences lead beyond the issue of time to another, closely related one: the problem of life. It is worth noting that Heidegger returns to this problem shortly after *Being and Time*, just as Bergson gives it his full attention after *Matter and Memory*, resulting in *Creative Evolution*.[14] In order to focus on time, I have not followed up on Heidegger's critical remarks concerning Bergson's concept of life. However, Heidegger's reference to Bergson's *élan vital* as one way to understand the double movement of temporalization suggests that we have more to learn about the relationship between these thinkers with respect to the issue of life. How was Heidegger's own interpretation of life influenced by Bergson? Is he correct that Bergson understands life as a mute, unidirectional flow? Are there insights into this phenomenon that Heidegger appropriates from Bergson? Does Bergson have anything of value to contribute to a contemporary phenomenology of life?[15] One of Bergson's broadest philosophical concerns, already evident in his early works, is the tendency of modern philosophy and science to reduce life—especially human life—to a set of mechanical, calculable operations. In this regard, there may be ways in which Bergson anticipates Heidegger's later thinking that deserve further exploration.

Today, when Bergson's thought is not ignored, there is a tendency to think of it as old-fashioned. His ideas are commonly regarded as passé, his theories as dated, and his methods as pre-phenomenological. This view is in accord with Heidegger's characterization of Bergson's approach to time, and his philosophy as a whole, as "traditional." However, to support his critique of Bergson, Heidegger arrests Bergson's thinking at its earliest stage, the theory of duration presented in *Time and Free Will*. I hope to have offered good reasons for concluding that this reading of Bergson misses the mark in some important ways, that Heidegger's own interpretation of temporality bears traces of a deep engagement with Bergson, and that Bergson's thinking on this issue is more radical than Heidegger is willing to admit. The way that Heidegger dismisses Bergson in *Being and Time* conceals a significant debt, or at least a productive, mostly behind-the-scenes collaboration. In the future, perhaps this debt will be repaid, but only if we can recall how Bergson's thinking helped clear the way for Heidegger's.

Notes

Introduction

1. See, for example, Leonard Lawlor, *The Challenge of Bergsonism: Phenomenology, Ontology, Ethics* (New York: Continuum, 2003) and "The Beginnings of Post-Modernism: Phenomenology and Bergsonism, Derrida and Deleuze," in *Thinking through French Philosophy: The Being of the Question* (Bloomington: Indiana University Press, 2003), 109–22. See also the excellent collection *Bergson and Phenomenology*, ed. Michael R. Kelly (London: Palgrave Macmillan, 2010).

2. Gilles Deleuze, *Cinema I: The Movement-Image*, trans. Hugh Tomlinson and Barbara Habberjam (Minneapolis: University of Minnesota Press, 1986), 219.

3. Maurice Merleau-Ponty, *In Praise of Philosophy and Other Essays*, trans. John Wild, James Edie, and John O'Neill (Evanston: Northwestern University Press, 1970), 15.

4. Emmanuel Levinas, *Ethics and Infinity: Conversations with Philippe Nemo*, trans. Richard A. Cohen (Pittsburgh: Duquesne University Press, 1985), 27.

5. Jacques Derrida, *Of Spirit: Heidegger and the Question*, trans. Geoffrey Bennington and Rachel Bowlby (Chicago: University of Chicago Press, 1989), 64.

6. As different as their ideas of philosophy are, Bergson and Heidegger agree that, as David Wood aptly puts it, "the question of time is not merely one of great philosophical interest as a topic for philosophical inquiry, but one that bears on the nature of that inquiry itself." David Wood, *The Deconstruction of Time* (Evanston: Northwestern University Press, 1989), 143.

7. See Edmund Husserl, *On the Phenomenology of the Consciousness of Internal Time (1893–1917)*, ed. Rudolf Boehm and trans. John B. Brough (Dordrecht: Kluwer Academic Publishers, 1991). The first edition of Husserl's lectures on time, delivered at the University of Göttingen in the winter of 1904–5, was published in 1928 in Husserl's journal, *Jahrbuch für Philosophie und phänomenologische Forschung*. The lectures had previously been edited by Husserl's assistant Edith Stein, but Husserl held them back and later allowed Heidegger to supervise their publication. See Brough's "Translator's Introduction," xi–xviii.

8. In 1913, Heidegger took Rickert's course "Seminar on Metaphysics in Conjunction with the Writings of H. Bergson." See Kisiel and Sheehan, *Becoming Heidegger: On the Trail of His Early Occasional Writings, 1910–1927*, ed. Theodore Kisiel and Thomas Sheehan (Evanston: Northwestern University Press, 2007), xxxviii.

9. For more on Bergson's life and philosophy, see Keith Ansell Pearson and John Mullarkey, eds., *Henri Bergson: Key Writings* (London: Continuum, 2002), viii–xi and 1–45; Suzanne Guerlac, *Thinking in Time: An Introduction to Henri Bergson* (Ithaca: Cornell University Press, 2006), 1–13; Gary Gutting, *French Philosophy in the Twentieth Century* (Cambridge: Cambridge University Press, 2001), 49–83; and Alan Schrift, *Twentieth Century French Philosophy: Key Themes and Thinkers* (Malden, MA: Blackwell Publishing, 2006), 99–101.

10. Husserl, *On the Phenomenology of the Consciousness of Internal Time*, 9–10.

Chapter 1

1. In §19 of *The Basic Problems of Phenomenology*, Heidegger provides his most detailed interpretation of Aristotle's account of time, including a critique of Bergson. See BPP 227–57/322–63.

2. Heidegger briefly discusses both Bergson's and Husserl's approaches to time in §12 of *The Metaphysical Foundations of Logic*. See MFL 203–4/262–4. Prior to *Being and Time*, he offers a critique of Hegel's interpretation of time in §20 and an examination of Aristotle's influence on Hegel and Bergson in §21 of *Logic: The Question of Truth*. See LQT 208–24. Among these recent thinkers (including Bergson), his deepest engagement is with Dilthey and his conception of history, especially in the 1925 Kassel lectures "Wilhelm Dilthey's Research and the Struggle for a Historical Worldview" (S 147–76) and the 1925 essay "The Concept of Time," recently published as *The Concept of Time: The First Draft of* Being and Time, trans. Ingo Farin with Alex Skinner (London: Continuum, 2011).

3. This is the same footnote that Derrida uses to show how and why Heidegger accuses Hegel of "thinking in terms of the vulgar concept of time." As Derrida notes, "it is by far the longest in *Being and Time*, pregnant with developments that are announced and held back, necessary but deferred." Jacques Derrida, "*Ousia* and *Grammē*: A Note on a Note from *Being and Time*," in *Margins of Philosophy*, trans. Alan Bass (Chicago: University of Chicago Press, 1982), 35. I hope to show that one of the necessary developments that were deferred was a decisive confrontation with Bergson's philosophy of time.

4. These were Bergson's principle thesis and his Latin thesis for his *doctorat*, which he submitted to the Université de Paris (Sorbonne) and published in 1889. The *Essai* was later translated into English as *Time and Free Will*, and *Quid Aristoteles de loco senserit* was translated by John K. Ryan as "Aristotle's Concept of Place," in *Ancients and Moderns: Studies in Philosophy and the History of Philosophy*, vol. 5 (Washington, DC: The Catholic University of America Press, 1970), 20–72;

and by Robert Mossé-Bastide as *L'idée de lieu chez Aristote*, in *Melanges*, ed. André Robinet (Paris: Presses Universitaires de France, 1972), 1–56.

5. See HCT 8–9/11–12 and LQT 206/249 and 339/411.

6. For an account of Heidegger's critique of Bergson in these courses, see Stephen Crocker, "The Oscillating Now: Heidegger on the Failure of Bergsonism," *Philosophy Today* 41 (1997): 405–23.

7. See Chapter 4 for a discussion of Bergson in *The Basic Problems of Phenomenology*.

8. Along with his thesis "The Theory of Categories and Meaning in Duns Scotus," this lecture was required for the *Habilitation* degree and the rank of *Privatdozent*. See Kisiel and Sheehan, *Becoming Heidegger*, xviii.

9. This project was largely inspired by Dilthey, who maintained that the human sciences should not proceed on the basis of the methods and assumptions of the natural sciences. See Wilhelm Dilthey, *Introduction to the Human Sciences*, ed. Rudolf A. Makkreel and Frithjof Rodi (Princeton: Princeton University Press, 1989). Among the disciplines included in the "human sciences" would be psychology, sociology, anthropology, politics, history, philosophy, literary studies, and art history. For more background on the term *Geisteswissenschaften*, see Charles R. Bambach, *Heidegger, Dilthey, and the Crisis of Historicism* (Ithaca: Cornell University Press, 1995), 128n. The question of the appropriate method for the study of human beings as historical beings was of great concern to the early Heidegger. A decade later, he identifies the common thread running through Dilthey's work in the history of science, epistemology, and psychology as "a vital questioning after the sense of history and human being" (S 154). This is a fitting description of many of Heidegger's own early works, including those that focus on the concept of time.

10. In contrast to the ontological problems that become central to Heidegger, here he focuses more on epistemological issues. As Kisiel says, this lecture "constitutes an early, and rare, venture on his part into the philosophy of the sciences, understood by him as a 'logical' deliberation on the sense . . . of their ground concepts or basic categories, which accordingly drives the initially epistemological analysis of a science toward exposure of its correlative ontological implications." Kisiel and Sheehan, *Becoming Heidegger*, xviii. That is, we can already see in the way that Heidegger deals with epistemological issues like concept formation a "tendency toward metaphysics" (S 49) leading in the direction of questions about the nature and existence—or, more properly, the being—of time.

11. Kisiel and Sheehan, *Becoming Heidegger*, xxxviii.

12. Dastur notes that "Heidegger, developing a theory of physical time which is not without similarities to Bergson's *Essai sur les données immédiates de la conscience* (cited by Heidegger in both of his theses), shares Dilthey's view of the difference between the respective methods of the natural and the human sciences." Françoise Dastur, *Heidegger and the Question of Time*, trans. François Raffoul and David Pettigrew (Atlantic Highlands, NJ: Humanities Press, 1998), 2.

13. Theodore Kisiel, *The Genesis of Heidegger's* Being and Time (Berkeley: University of California Press, 1993), 16–17.

14. Although the essay is ostensibly a book review, according to Kisiel, "The review is thus as much a 'discourse on method' of Heidegger's own devising as it is about the substance of Jaspers's book." Kisiel, *Genesis*, 139.

15. This distinction plays a prominent role in Heidegger's 1920 summer course *Phenomenology of Intuition and Expression*, which I discuss below.

16. Natorp poses the problem in this way: "If one were to try, if it were at all possible, to somehow grasp the content of immediate experience purely as it is in itself . . . would one not be forced to artificially still and interrupt the continuous stream of becoming, which surely is how inner life presents itself?" Paul Natorp, *Allgemeine Psychologie nach kritischer Methode* (Tubingen: J. C. B. Mohr, 1912), 102–3; quoted in Dan Zahavi, "How to Investigate Subjectivity: Natorp and Heidegger on Reflection," *Continental Philosophy Review* 36 (2003): 157. Heidegger offers an extensive critical discussion of Natorp's philosophy in his 1920 summer course. See PIE 73/92ff.

17. The view of lived experience that Heidegger attacks here is one that he considers to be at the root of a number of debates between phenomenologists (including Husserl), neo-Kantians, and others. He finds the alternatives to "rationalist" trends in philosophy unsatisfying because, as Fehér writes, "epistemologically oriented philosophy fails to seize upon life because its outlook and conceptuality are rooted in theoretical comportment, and in a derivative mode of it. But the opponents, life-philosophy, historicism, and any kind of irrationalism, remain dependent upon it precisely to the extent to which they claim life, history, or existence to be inaccessible to concepts." István M. Fehér, "Phenomenology, Hermeneutics, *Lebensphilosophie*: Heidegger's Confrontation with Husserl, Dilthey, and Jaspers," in *Reading Heidegger from the Start: Essays in His Earliest Thought*, ed. Theodore Kisiel and John Van Buren (Albany: State University of New York Press, 1994), 87. One of Heidegger's main aims in discussing contemporary philosophers of life, among whom he includes Bergson, is to draw attention to the implicit privileging of the theoretical attitude common to both "rationalism" and "irrationalism."

18. On Bergson's method of philosophical intuition, see CM 30–90/1271–330, 107–29/1345–65, and 159–200/1392–432.

19. Heidegger offers a positive account of the modification of worldly or practical dealings into theoretical apprehension in §40 of *Being and Time*. See BT 340–46/356–64.

20. Heidegger indicates his continuing interest in Bergson with his use of a quotation from *Matter and Memory* as an epigraph to the course: "*nous sommes en* [*sic*] *d'ouvrir toujours devant nous l'espace, de refermer toujours derrière nous la durée* [we are always opening space out before us, always closing duration off behind us]" (BPPa 1/1). See MM 148–9/290.

21. The others occur in *History of the Concept of Time* (SS 1925), *Logic: The Question of Truth* (WS 1925–6), and *Being and Time*.

22. In summary fashion, Heidegger writes, "*Simmel, Bergson, James:* Intuition of life—'creative evolution [*schöpferische Entwicklung*]'—pragmatism" (BPPa 8/19).

Although he mentions *Creative Evolution* elsewhere (see BPPa 216/108n and BPPb 232/328) and occasionally refers to Bergson's *élan vital* (see TDP 88/115 and MFL 193/249, 207/268), the critique of Bergson that Heidegger develops during this period is based entirely on *Time and Free Will* (see PIE 19/26, HCT 23/28, LQT 221/267, and especially BT 410n/432n).

23. This comment suggests that Heidegger might have Bergson in mind in when he begins his 1919 lecture "On the Nature of the University and Academic Study": "*Situation in the life-context:* a situation is a certain unity in natural experience [*Erlebnis*]. Situations can interpenetrate one another: their durations do not exclude each other (e.g., a year in the field, a semester: no objective concept of time)" (TDP 153/207). In a lecture on this topic in his course "Introduction to the Phenomenology of Religion" (WS 1920–1), his reference is more explicit: "'Situation' . . . does not designate anything in the manner of an order. . . . A situational series is not, moreover, a series in the manner of an order (compare to Bergson's '*durée concrète*')." Martin Heidegger, *The Phenomenology of Religious Life*, trans. Matthias Fritsch and Jennifer Anna Gosetti-Ferencei (Bloomington: Indiana University Press, 2010), 63.

24. On two other occasions in the notes and transcripts for this course, Heidegger mentions Bergson while criticizing Oswald Spengler's *Decline of the West*: "Its conceptions, like those of Bergson, Dilthey, and Simmel, are to be understood as not original and as not necessarily pertaining to what is radical, and so dismissed" (BPPa 123/160); and "*Spengler*, just like *Bergson, Dilthey, Simmel*, is bogged down in half-measures. He has not apprehended life back up to its ultimacy" (BPPa 190/253). These are the first occasions where Heidegger employs a strategy of establishing guilt by association with other philosophers of life, as he does later in *History of the Concept of Time* and *Being and Time.*

25. Concerning the role of "The Concept of Time" in the development of *Being and Time*, Kisiel writes: "Gadamer aptly calls it the '*Urform*' of BT. Not yet a draft of BT, it is nevertheless the first major and quite public step toward the extant book BT, by elaborating its core structure, which in retrospect finds its seminal roots in the 1922 Introduction [to Aristotle's philosophy]." Kisiel, *Genesis*, 315. See also Thomas Sheehan, "The 'Original Form' of *Sein und Zeit*: Heidegger's *Der Begriff der Zeit* (1924)," *Journal of the British Society for Phenomenology* 10 (1979): 78–83; Theodore Kisiel, "Why the First Draft of *Being and Time* Was Never Published," *Journal of the British Society for Phenomenology* 20 (1989): 3–22; and John Sallis, "Time Out . . . ," in *Echoes: After Heidegger* (Bloomington: Indiana University Press: 1990), 44–75.

26. This passage and the discussion following it anticipate the analyses of time measurement in *Being and Time*, §80 (BT 391–9/465–72/413–20) and clock usage in *The Basic Problems of Phenomenology*, §19 (BPP 257–61/363–9). Heidegger's analyses are examined against the background of the confrontation between Bergson and Einstein in David Scott, "The 'Concept of Time' and the 'Being of the Clock': Bergson, Einstein, Heidegger, and the Interrogation of the Temporality of Modernism," *Continental Philosophy Review* 39 (2006): 83–213.

27. The term *Jeweiligkeit* is also prominent in Heidegger's 1923 summer course. In the notes to his translation of that course, Van Buren explains: "The neologism *Jeweiligkeit* is coined from the adjective *jeweilig*, which normally means 'respective,' 'prevailing,' or 'at the particular time' (cf. *der jeweilige König* [the king at the particular time]), but has the literal meaning of 'in each case [*je*] for a while at the particular time [*weilig*].'' Accordingly, in coining the term *Jeweiligkeit*, Heidegger has in mind both its more conventional meaning of something like 'temporal particularity' and its literal meaning of 'awhileness.' . . ." Martin Heidegger, *Ontology: The Hermeneutics of Facticity*, trans. John van Buren (Bloomington: Indiana University Press, 1999), 108. See also William McNeill's explanation in his translation of "The Concept of Time" (CT 24–25, n. 8).

28. When Heidegger identifies Dasein with time, he takes a big step in Bergson's direction. Merleau-Ponty finds something similar in Bergson, who recognizes "that we do not draw near to time by squeezing it between the reference-points of measurement as if between pincers; but that in order to have an idea of it we must let it develop freely. . . . from without, I would only have the trail of time; I would not be present in its generative thrust. So time is myself; I am the duration I grasp, and time is duration grasped in me." Maurice Merleau-Ponty, *Signs*, trans. Richard C. McCleary (Evanston: Northwestern University Press, 1964), 184.

29. Heidegger drops the term *Vorbei* from his analysis in *Being and Time*, replacing it with *Gewesenheit* (translated as "having-been" or "beenness"). See Sallis, "Time Out . . . ," 51–52.

30. The origin of this concept is the issue of *Being and Time*, §81. See BT 472–80/420–8. See also BT 39/17–18, 48–49/26, and 500–1/432–3n.

31. In *Being and Time*, Heidegger replaces the characterization of time as "homogeneous" with the expression "leveled down." There, he writes: "In the vulgar interpretations of time as a succession of nows . . . the ecstatic and horizonal constitution of temporality . . . is *leveled down* [*nivelliert*]. . . . This covering over and leveling down of world time that is carried out by the vulgar understanding of time is no accident" (BT 401/474/422). This terminological shift effectively conceals the proximity of Heidegger's thinking about the way time is commonly understood or represented to Bergson's.

32. Martin Heidegger, *The Concept of Time: The First Draft of* Being and Time, op. cit. Heidegger does not mention Bergson in this article, but evokes him nevertheless in comments like this one: "reckoning with time never turns it into 'space.' Time cannot be spatialized. . . . the calculational approach to time is a particular kind of *temporalization* [*Verzeitlichen*] in the mode of presencing [*Gegenwärtigen*]" (68). Heidegger fleshes this out in BT §80, which I discuss in Chapter 3.

33. See MFL 203/262.

34. Heidegger himself attributes this theme to Bergson. The preliminary part of *History of the Concept of Time* on "The Sense and Task of Phenomenological Research" is not repeated in *Being and Time*, but division one of the main part (comprising the second half of the course), "Preparatory Description of the Field in Which the Phenomenon of Time Becomes Manifest," is a rough draft of division

one of *Being and Time*. On the connection between these two works, see Theodore Kisiel, "On the Way to *Being and Time*: Introduction to the Translation of Heidegger's *Prolegomena zur Geschichte des Zeitbegriffs*," *Research in Phenomenology* 15 (1985): 193–226, and *The Genesis of Heidegger's* Being and Time, 363–97.

35. During this summer course, Heidegger only completes the preliminary introduction to phenomenology and the first division of the first part, "The preparatory description of the field in which the phenomenon of time becomes manifest" (HCT 8/11). As a result of this extensive preparatory work, he barely touches on the issue of time until the final session of the course, when he anticipates the interpretation of temporality as the meaning of care that will be worked out in *Being and Time*. As he reenvisions the project in his outline for *Being and Time*, the phenomenological destruction will begin with Kant, turn next to Descartes, and conclude with Aristotle (BT 63–64/39–40). With that, Bergson drops out of Heidegger's plan for showing how the concept of time arises and gets handed down through the philosophical tradition.

36. This claim that Brentano influenced Bergson is intriguing because Bergson does not regard consciousness, as phenomenologists do, as defined by intentionality. For Deleuze, this marks a fundamental difference between Bergson and phenomenologists: "It was necessary, at any cost, to overcome this duality of image and movement, of consciousness and thing. Two very different authors were to undertake this task at about the same time: Bergson and Husserl. Each had his own war cry: all consciousness is consciousness *of* something (Husserl), or more strongly, all consciousness *is* something (Bergson)." Deleuze, *Cinema I*, 56.

37. Husserl's lectures on time-consciousness at the University of Göttingen had been delivered almost a decade earlier, in 1904–5, so it is unclear what impact Heidegger believes Bergson could have had on Husserl's thinking, unless he means to suggest an influence on the 1917–18 "Bernau Manuscripts." See Edmund Husserl, *Die 'Bernauer Manuskripte' über das Zeitbewußtsein (1917/18)*, Husserliana XXXIII, ed. Rudolf Bernet and Dieter Lohman (Dordrecht: Kluwer Academic Publishers, 2001). Thanks to Michael Kelly for bringing this question to my attention.

38. See Edmund Husserl, *Ideas Pertaining to a Pure Phenomenology and to a Phenomenological Philosophy, First Book: General Introduction to a Pure Phenomenology*, trans. Fred Kersten (The Hague: Kluwer Academic Publishers, 1982).

39. See Edmund Husserl, "Philosophy as a Rigorous Science," trans. Quentin Lauer, in *Phenomenology and the Crisis of Philosophy* (New York: Harper and Row, 1965), 71–147.

40. See Edmund Husserl, *Ideas Pertaining to a Pure Phenomenology and to a Phenomenological Philosophy, Second Book: Studies in the Phenomenology of Constitution*, trans. Richard Rojcewicz and André Schuwer (The Hague: Kluwer Academic Publishers, 1989).

41. See Max Scheler, *Formalism in Ethics and Non-formal Ethics of Values: A New Attempt Toward the Foundation of an Ethical Personalism*, trans. Manfred S. Frings and Roger L. Funk (Evanston: Northwestern University Press, 1973). Heidegger's critique focuses on part two, chapter six, "Formalism and Person," but he

also refers to Scheler's "The Idols of Self-Knowledge," trans. David R. Lachterman, in *Selected Philosophical Essays* (Evanston: Northwestern University Press, 1973) 3–97.

42. See *Being and Time* §12 for a corresponding discussion of being-in that mentions spatiality but not temporality (or "whileness").

43. The task of phenomenological chronology, as Dastur explains it, is "to investigate the temporality of phenomena, by which must be understood not their being in time, which still remains an extrinsic determination, but what in their very structure is characterized by time." Françoise Dastur, *Telling Time: Sketch of a Phenomenological Chrono-logy*, trans. Edward Bullard (London: Athlone Press, 2000), 8.

44. Hegel's discussion of time appears in the second part of the *Encyclopaedia*, translated by A. V. Miller as *Hegel's Philosophy of Nature* (Oxford: Clarendon Press, 1970). Kant discusses time in several well-known sections of his *Critique of Pure Reason*, trans. Norman Kemp Smith (New York: St. Martin's Press, 1929). He deals with time first in the Transcendental Aesthetic (A19–49/B33–73) and later in the Analogies of Experience (A177–218/B218–65), but Heidegger is most interested in how the problem of time shows up in the Schematism (A137–47/B176–87). See §§22–36 of LQT and §§32–35 of Martin Heidegger, *Kant and the Problem of Metaphysics*, trans. Richard Taft (Bloomington: Indiana University Press, 2000).

45. Sheehan notes in his translation of this course that Heidegger writes "succession" in French (rather than the German *Sukzession*, *Folge*, or *Nacheinander*). Also, the term *Temporal* is rendered as "ur-temporal" throughout Sheehan's translation, but in this passage he uses "ur-time" for *ursprünglichen Zeit*, which other translations of Heidegger's early works render as either "primordial time" or "originary time."

46. The italicized portion reads "*da is die Zeit als Jetzt-Zeit verstanden, und zeitlich besagt: 'in die Zeit' fallend, 'in der Zeit' ablaufend.*" The language Heidegger uses here points to Hegel and his description of Spirit "falling [*fallend*] into time," as well as to Husserl and his analysis of the "running-off" [*ablaufend*] phenomena of retentional consciousness. See §82 of BT and §10 of Husserl, *On the Phenomenology of the Consciousness of Internal Time.*

47. Compare BT 410n/432–3n.

48. I return to Heidegger's evidence for Bergson's dependence on Aristotle and offer my evaluation of it in Chapter 2.

49. See TFW 91/62. Heidegger translates Bergson's *milieu* as "field" (*Feld*), while Pogson uses the more literal "medium" in his English translation.

50. See BT xxix/1, 17–18/17–18, 24–25/25–26, and 225/235.

51. Heidegger repeats these criticisms in §65 of BT and expands on them in §80 in the course of his discussion of the measurement of time and the "within-timeness" or "intratemporality" (*Innerzeitigkeit*) of things encountered in the world. See BT 318/333 and 397–8/418–19. See Chapter 3 for my reading of these sections.

Chapter 2

1. Heidegger's critique of Bergson foreshadows his treatment of Nietzsche in the 1930s, as Leonard Lawlor explains: "In the most famous footnote from *Being*

and Time—made famous obviously by Derrida—Heidegger claims that Bergson merely 'reverses' Aristotle's views on time. Heidegger's use of the word 'reversal' here anticipates its use in para. 24 of his first set of Nietzsche lectures, where he claims that Nietzsche merely 'reverses' Platonism without 'twisting free' of it." Leonard Lawlor, *The Challenge of Bergsonism*, 28.

2. See the first editor's note to Bergson's *Œuvres*, 1485.

3. Bergson, "Aristotle's Concept of Place," 20.

4. Bergson writes:

> [T]he multiplicity of conscious states, regarded in its original purity, is not at all like the discrete multiplicity which goes to form a number. In such a case there is, as we have said, a qualitative multiplicity. In short, we must admit two kinds of multiplicity, two possible senses of the word "distinguish," two conceptions, the one qualitative and the other quantitative, of the difference between *same* and *other*. Sometimes this multiplicity, this distinctness, this heterogeneity contains number only potentially, as Aristotle would have said. Consciousness, then, makes a qualitative discrimination without any further thought of counting the qualities or even of distinguishing them as *several*. In such cases we have a multiplicity without a quantity (TFW 121–2/81).

Interestingly, the sole reference to Aristotle in *Time and Free Will* is made not to contradict his definition of time, but to help clarify the relationship between the two multiplicities: duration, as qualitative and heterogeneous, "contains number only potentially."

5. I return to this issue in Chapter 4 when I examine Heidegger's critique of Bergson's alleged interpretation of Aristotle in *The Basic Problems of Phenomenology* (SS 1927).

6. In *Being and Time* Heidegger lays emphasis on Bergson's use of the terms *quantitative* and *qualitative*, evidence of Aristotle's influence, in his view, but in his 1924–5 *Logic* course he seems even more concerned about how Bergson identifies duration with succession: "A clear indication that Bergson failed to break through to a conception and categorical knowledge of ur-time is the fact that he understood even lived time—duration—as (in French) '*succession*,' with the sole proviso that the *succession* of lived time is, he says, not a quantitative *succession* laid out in individual now-points. Instead, this *succession* is a qualitative one, in which the individual moments of time—past, present, and future—permeate one another. With that he has reached the limit, for he does not say what quantity or quality is; he provides no principled discussion of these two guiding threads, but simply presupposes them as already known. Moreover, he describes qualitative time—duration—merely in images, with not a word about working out any kind of concept. The essential point, therefore, is that Bergson certainly does try to get to authentic time by way of the phenomenon of duration, but he understands duration once again in terms of *succession*" (LQT 207/249–50). This shows that it is not only the employment of Aristotelian categories that makes Bergsonian duration "traditional," for Heidegger,

but also its conformity to the concept of time as a "succession of nows," which is the theme of §81 of *Being and Time*. To ignore Bergson's distinction between the two multiplicities as Heidegger does is to risk misinterpreting what he means by "succession."

7. Deleuze highlights "temporalizing," or "thinking in terms of duration," as one of three aspects of Bergson's method. See *Bergsonism*, 31–35.

8. The challenge is to think, as Ansell Pearson puts it, "beyond our dominant habits of representation in which time is conceived in terms of space." Keith Ansell Pearson, *Philosophy and the Adventure of the Virtual: Bergson and the Time of Life* (London: Routledge, 2002), 10. This means not only criticizing the way certain philosophical problems are posed, but also creating new ways of thinking about the issue. In this regard, Deleuze writes, "the effort of invention consists most often in raising the problem, in creating the terms in which it will be stated." Deleuze, *Bergsonism*, 15.

9. See Kant, *Critique of Pure Reason*, A22–36/B37–53.

10. See Deleuze, *Bergsonism*, 112–13.

11. In his 1925–6 *Logic* course, Heidegger writes, "Bergson's goal is to work out the difference between time and duration. Duration, however, for Bergson is nothing but lived time . . . considered in the way it shows up in consciousness" (LQT 207/249).

12. Bergson's title, *Essai sur les données immédiates de la conscience*, announces not so much his subject matter as his method: to view philosophical problems in the light of the "immediate data of consciousness." Translating *données* more literally, we could say "what is immediately given to consciousness." This title, plus Bergson's choice of the term "intuition" for his method, has invited numerous comparisons with phenomenology, especially Husserlian phenomenology. Bergson's method of intuition does not become a theme of his philosophy until after *Time and Free Will*, but he already employs it there. Most of his subsequent works include discussions of this method. Bergson's most detailed elaborations on his method appear in the second introduction to *The Creative Mind* (CM 30–90/1271–330) and, in that same volume, the essays "Philosophical Intuition" (CM 107–29/1345–65) and "Introduction to Metaphysics" (CM 159–200/1392–432). For more on Bergsonian intuition, see Deleuze, *Bergsonism*, 13–35 and Lawlor, *Early Twentieth-Century Continental Philosophy* (Bloomington: Indiana University Press, 2012), 15–37.

13. It may seem inappropriate to talk about Bergson's "concept of duration" because he denies that the experience of duration can be grasped conceptually. In "Introduction to Metaphysics," for instance, he argues that duration "cannot be represented by images. But still less could it be represented by *concepts*, that is, by abstract ideas, whether general or simple" (CM 165/1399). However, he does not reject conceptual understanding as such, but only "fixed" concepts. He argues that metaphysics "is strictly itself only when it goes beyond the concept, or at least when it frees itself of the inflexible and ready-made concepts and creates very different ones from those we usually handle, I mean, flexible, mobile, almost fluid representations, always ready to mold themselves on the fleeting forms of intuition"

(CM 168/1401–2). Bergson can therefore be understood as attempting to create *new concepts* and devise new ways of thinking that are more precise and better fitted to experience. See Gilles Deleuze, "Bergson's Conception of Difference," in *Desert Islands and Other Texts (1953–1974)*, ed. David Lapoujade, trans. Michael Taormina (Los Angeles: Semiotext(e), 2004), 33. See also Deleuze, *Bergsonism*, 44–45 and Lawlor, *Early Twentieth-Century Continental Philosophy*, 30ff.

14. Later, Bergson reconsiders whether psychological states, particularly sensations, are extended. In *Matter and Memory*, he argues that they are, and that their extension makes them continuous with the qualities of material objects, but he continues to insist that they be distinguished from anything juxtaposed in space. See MM 55–63/203–11, 139–41/281–3, and 212–18/346–52.

15. In one of these experiments, performed by Belgian psychologist J. R. L Delbœuf, subjects were instructed to view three concentric rings of light and, while the researchers increased and decreased the light, to indicate whether the difference in brightness between the first and second rings was equal to the difference between the second and third (TFW 56/42). The subjects' ability to do this with some accuracy led Delbœuf to conclude that the sensation of brightness is a measurable quantity, and he sought to construct a "scale of luminous intensities" by which that kind of sensation can be measured. Bergson argues that as the rings of light grow brighter or dimmer, all that the subjects immediately experience are series of qualitative changes, but that thanks to an acquired skill, they can estimate when the difference in luminosity between the light sources of the first two rings is equal to the difference between those of the second two. Thus, the subjects really measure the intensity of the light, not the sensation of brightness—that is, they measure the physical cause, not the psychological effect. On Bergson's critique of psychophysics, see Guerlac, *Thinking in Time*, 24–25; and John Mullarkey, *Bergson and Philosophy* (Notre Dame, IN: University of Notre Dame Press, 1999), 22–24.

16. In either case, the consequences of failing to distinguish between qualitative progress and quantitative increase are the same. As Deleuze explains: "Whether the quality of the sensation is confused with the muscular space that corresponds to it, or with the quantity of the physical cause that produces it, the notion of intensity involves an impure mixture between determinations that differ in kind, so that the question 'by how much does the sensation grow?' always goes back to a badly stated problem." Deleuze, *Bergsonism*, 18–19. In Chapter 5, I show that the question "where are memories stored?" receives a similar criticism by Bergson in *Matter and Memory*: like the question about the intensity of sensations, it betrays a habit of spatial thinking.

17. Bergson's analysis of intensity anticipates Merleau-Ponty's early phenomenology in this respect. On the one hand, Merleau-Ponty is sharply critical of Bergson: "This phenomenal field is not an 'inner world,' the 'phenomenon' is not a 'state of consciousness,' or a 'mental fact,' and the experience of phenomena is not an act of introspection or intuition in Bergson's sense. ...The return to 'the immediate data of consciousness' [was] a hopeless enterprise." Maurice Merleau-Ponty, *Phenomenology of Perception*, trans. Colin Smith (London: Routledge, 1962), 66. On

the other hand, Merleau-Ponty follows Bergson in rejecting the notion that there are objective, causal relationships between sensations and feelings. See, for example, the brief discussion of Proust's *Remembrance of Things Past* in the temporality chapter of *Phenomenology of Perception*, 493–94. For more on the relation between Merleau-Ponty's thought and Bergson's, see Alia Al-Saji, "The Temporality of Life: Merleau-Ponty, Bergson, and the Immemorial Past," *The Southern Journal of Philosophy* 45 (2007): 177–206; Ted Toadvine, "Nature and Negation: Merleau-Ponty's Reading of Bergson," *Chiasmi International* 2 (2000): 107–118; and Dorothea Olkowski, "Merleau-Ponty and Bergson: The Character of the Phenomenal Field," in *Merleau-Ponty: Difference, Materiality, Painting*, ed. Véronique M. Fóti (Atlantic Highlands, NJ: Humanities Press, 1996), 27–36.

18. This explains why Bergson goes on to characterize duration as "qualitative multiplicity." Without taking into account his discussion of intensity, it may be unclear what Bergson means by this expression. Or, as William Blattner puts it, "It is a little hard to know just what Bergson is driving at, because his argument for the purely qualitative character of time falls so woefully short of its conclusion." William Blattner, *Heidegger's Temporal Idealism* (Cambridge: Cambridge University Press, 1999), 268. With that discussion in mind, however, we can see that Bergson's claim is not that that the parts of time—past, present, and future—are qualitatively different, but rather that our conscious experience is qualitatively different at each moment. For that reason, the moments of time do not just follow one another, but each moment is qualitatively distinct from the last. Bergson clarifies this in chapter two of *Time and Free Will* with several illustrations of the difference between counting and experiencing qualitative change, such as the example of a swinging pendulum. See TFW 104–6/70–71.

19. On Bergson's distinction between these two forms of multiplicity, see Deleuze, *Bergsonism*, 37–49; Ansell-Pearson, *Philosophy and the Adventure of the Virtual*, 13–24; and Robin Durie, "Splitting Time: Bergson's Philosophical Legacy," *Philosophy Today* 44 (2000): 154–8. The term "multiplicity" and the distinction are borrowed from the mathematician Bernhard Riemann. The French '*multiplicité*' corresponds to the German '*Mannigfältigkeit*' and may also be translated as "manifoldness."

20. In this respect, Bergson's thinking is in accord with James's. See William James, *Principles of Psychology*, vol. 1 (New York: Henry Holt & Co., 1890), 605–42.

21. Bergson's view of space may strike some as Newtonian rather than Kantian, to the extent that he considers space "a homogeneous and immobile container wholly unaffected by the material bodies and movements it supposedly contains." Mullarkey, *Bergson and Philosophy*, 12. However, as Mullarkey explains, "There is a historical development in Bergson's understanding of space, with a positive conception of it emerging to counteract the negative presentations that are mostly confined to his first book. By the time Bergson publishes *MM*, what appeared as a real property of space in *TFW*'s depiction [i.e., homogeneity] has become the product of our pragmatic interaction with it" (12–13). By emphasizing the role of the body in perception and the utilitarian character of our concept of space, Bergson indicates

the difference between homogeneous, absolute space and the spatiality of the lived body from which it arises.

22. In his essay "Introduction to Metaphysics," Bergson uses three different images—an unrolling spool, a color spectrum, and a piece of elastic—to illustrate three features of duration—its continuity, heterogeneity, and indivisibility (CM 164–5/1397–8). For an elucidation of these images, see Lawlor, *Early Twentieth-Century Continental Philosophy*, 20–21.

23. The experience of hearing a melody is also a privileged example for Husserl, who uses it to illustrate the *a priori* necessity of retentional awareness. See, for example, Husserl, *On the Phenomenology of the Consciousness of Internal Time*, 22ff.

24. Bergson later makes a similar argument on the basis of the experience of hearing a clock chime: "Whilst I am writing these lines, the hour strikes on a neighboring clock, but my inattentive ear does not perceive it until several strokes have made themselves heard. Hence I have *not* counted them; and yet I only have to turn my attention backwards to count up the four strokes which have already sounded and add them to those which I hear. . . . I perceive that the first four sounds had struck my ear and even affected my consciousness, but that the sensations produced by each one of them, instead of being set side by side, had melted into one another in such a way as to give the whole a peculiar quality. . . . In a word, the number of strokes was perceived as a quality and not as a quantity: it is thus that duration is presented to immediate consciousness" (TFW 127–8/94–95). According to this description, not just the measurement of time, but all counting implies an original synthetic organization in consciousness.

25. The sharp distinction that Bergson makes here between states of consciousness (succession without mutual externality) and objects in space (mutual externality without succession) is one that he revisits in *Matter and Memory*. For, with this distinction, "[h]is thinking is now forced to find ways of overcoming the fundamental antinomy that has been set in place between mind and world." Ansell Pearson, *Philosophy and the Adventure of the Virtual*, 35. For a discussion of how the dualism of *Time and Free Will* helps set the agenda for *Matter and Memory*, see Jean Hyppolite, "Various Aspects of Memory in Bergson," trans. Athena V. Colman, in Lawlor, *The Challenge of Bergsonism*, 112–27.

26. Bergson discusses Zeno's paradoxes in a number of other works, because they provide the most vivid illustration of the kind of problems that arise when we replace real change and movement with artificial reconstructions. See, for example, MM 191–3/326–9 and CM 142–5/1377–80.

27. Later, Bergson writes: "Time could be enormously and even infinitely accelerated; nothing would be changed for the mathematician, for the physicist or for the astronomer. And yet the difference with regard to consciousness would be profound" (CM 13/1255).

28. In "Bergson's Conception of Difference," Deleuze highlights the heterogeneity of duration, which he defines as "internal difference" or "*what differs from itself*." Deleuze, *Desert Islands*, 37. In contrast to Heidegger, Deleuze interprets Bergsonism as a philosophy of difference and Bergsonian intuition as a "method of

division." See Deleuze, *Bergsonism*, 21–30. Also against Heidegger, Deleuze emphasizes "Bergsonism's incompatibility with Hegelianism, indeed with any dialectical method" (44). See also *Desert Islands*, 38–39 and 41ff.

29. In his 1924–5 *Logic* course, Heidegger actually displays a keen awareness of the difference, for Bergson, between time as it is *lived* and time as it is represented: "Bergson's goal is to work out the difference between time and duration. Duration, however, for Bergson, is nothing but lived time, and this lived time, in turn, is merely object-time or world-time, insofar as it is considered in the way it shows up in consciousness" (LQT 207/249). This remark is immediately followed by Heidegger's argument that Bergson fails to see "ur-time" because he identifies duration with succession.

30. Merleau-Ponty regards this as one of Bergson's earliest innovations: "In 1889 it was a great novelty—and one which had a future—to present as the basis of philosophy not an *I think* and its immanent thoughts but a Being-self whose self-cohesion is also a tearing away from self." Merleau-Ponty, *Signs*, 184.

31. Even Bergson admits that his descriptions are not free of spatial thinking: "I said that several conscious states are organized into a whole, permeate one another, gradually gain a richer content, and might thus give any one ignorant of space the feeling of pure duration; but the very use of the word 'several' shows that I had already isolated these states, externalized them in relation to one another, and, in a word, set them side by side: thus, by the very language which I was compelled to use, I betrayed the deeply ingrained habit of setting out time in space" (TFW 122/91).

32. Bergson explores the nature of dreams further in *Matter and Memory* (MM 154–5/295–6, 167–8/306–7 and 174/312–13) and a lecture titled "Dreams," in Henri Bergson, *Mind-Energy*, trans. H. Wildon Carr (New York: Henry Holt & Co., 1920), 104–33, which was also translated by Wade Baskin as *The World of Dreams* (New York: Philosophical Library, 1958).

33. In *Creative Evolution*, Bergson thinks about the difference between living organisms and isolated systems of matter in temporal terms: "The present state of an unorganized body depends exclusively on what happened at the previous instant," whereas in an organized body, "*all* the past of the organism must be added to that moment, its heredity—in fact, the whole of a very long history. . . . to know a living being or *natural system* is to get at the very interval of duration, while the knowledge of an *artificial* or *mathematical system* applies only to the extremity" (CE 19–23/510–13).

34. Bergson argues that philosophy has traditionally been incapable of thinking the radically new and unforeseeable because it suffers from the illusion that possibility precedes reality, whereas "reality casts its shadow behind it into the indefinitely distant past" (CM 22/1264). Because we can see the antecedents or precursors of future events in retrospect, we conclude that future events must have preexisted their realization. Bergson calls this the "retrograde movement of the true." In "The Possible and the Real," he claims, "the possible is only the real with the addition of an act of mind which throws the image back into the past" (CM 99–100/1339).

35. In "The Perception of Change" and "Philosophical Intuition," Bergson modifies Spinoza's formula *sub specie aeternitatis* to describe his own standpoint: "the more we accustom ourselves to think and to perceive all things *sub specie durationis*, the more we plunge into real duration" (CM 157–8; see also CM 128–9).

36. In *Creative Evolution*, to illustrate the difference between abstract time and concrete duration, Bergson considers the experience of waiting for sugar to melt in a glass of water: "For here the time I have to wait is not that mathematical time which would apply equally well to the entire history of the material world, even if that history were spread out instantaneously in space. It coincides with my impatience, that is to say, with a certain portion of my duration, which I cannot protract or contract as I like. It is no longer something *thought*, it is something *lived*" (CE 9–10/502).

37. As Kolakowski aptly puts it, "our self is at every moment, as it were, in a state of being born, absorbing its past and creating its future." Leszek Kolakowski, *Bergson* (South Bend, IN: St. Augustine's Press, 2001), 21.

38. This does not mean that Bergson thinks of freedom as effortless. See Hanne Jacobs and Trevor Perri, "Intuition and Freedom: Bergson, Husserl, and the Movement of Philosophy," in *Bergson and Phenomenology*, 101–17.

39. See Kant, *Critique of Pure Reason*, A445–7/B473–5 and A532–57/B560–85.

40. In *The Metaphysical Foundations of Logic*, during a discussion of ecstatic temporality, Heidegger echoes Bergson: "Only seldom do we take possession of time . . . only seldom do we become master of this power which we ourselves are; only seldom do we exist freely" (MFL 199/257–8).

Chapter 3

1. Aristotle, Augustine, and Kant are particularly important for Heidegger with respect to the issue of time. In *The Basic Problems of Phenomenology*, he also mentions Plotinus, Simplicius, Aquinas, Suarez, Leibniz, and Hegel (BPPb 231/328). He provides a similar compendium of classics on the problem of time in *The Metaphysical Foundations of Logic*, adding Husserl to the mix (MFL 198–9/256). Earlier in that same course, he mentions Kierkegaard, Dilthey, and Spengler as well (MFL 141/177–8). Heidegger's interpretation of temporality in *Being and Time* displays the influence of many if not most of these thinkers.

2. My understanding of temporality in *Being and Time* and connected works has benefited greatly from the following critical examinations: David Wood, *Time After Time* (Bloomington: Indiana University Press, 2007) and *The Deconstruction of Time*; Carol White, *Time and Death: Heidegger's Analysis of Finitude* (Burlington, VT: Ashgate Publishing Company, 2005); Tina Chanter, *Time, Death, and the Feminine: Levinas with Heidegger* (Stanford: Stanford University Press, 2001); Karin de Boer, *Thinking in the Light of Time* (Albany: State University of New York Press, 2000);

Françoise Dastur, *Telling Time: Sketch of a Phenomenological Chrono-Logy*, trans. Edward Bullard (London: Athlone Press, 2000) and *Heidegger and the Question of Time*; William Blattner, *Heidegger's Temporal Idealism*; Theodore Kisiel, *The Genesis and Structure of* Being and Time; John Sallis, *Echoes: after Heidegger*; Otto Pöggeler, *Martin Heidegger's Path of Thinking*, trans. Daniel Magurshak and Sigmund Barber (Atlantic Highlands, NJ: Humanities Press International, 1987); and Charles Sherover, *The Human Experience of Time: The Development of Its Philosophic Meaning* (Evanston: Northwestern University Press, 1975) and *Heidegger, Kant, and Time* (Bloomington: Indiana University Press, 1971).

3. The fact that the project of *Being and Time* is not complete by the end of division two is clear from its closing lines: "Is there a way leading from primordial *time* to the meaning of *being*? Does *time* itself reveal itself as the horizon of *being*?" (BT 415/437). These questions mark the transition from Heidegger's temporal interpretation of Dasein, grounding the existential structures discovered in division one, to a temporal interpretation of being as such. On Heidegger's efforts to follow through with the latter, see Dastur, *Heidegger and the Question of Time*, 53–69.

4. What is sometimes called the "privilege of presence" is, for Heidegger, a thread connecting ancient and modern philosophy. Françoise Dastur explains: "This fundamental question concerning the condition of the possibility of the understanding of Being by the Greek philosophers and their descendants, the modern philosophers of the West, as *constant presence*, is a question bearing on the relation—never made explicit in this Western tradition—between time and Being. What makes possible the understanding of being from a specific dimension of time, that is, the present? This is the question, in its most basic form, which is at the origin of *Being and Time*." Dastur, *Heidegger and the Question of Being*, xxiii. Heidegger views the tendency to understand being in terms of presence not only as a defining feature of Western philosophy, but also as a source of misunderstanding with respect to time. As Otto Pöggler puts it: "But wherein lies the actual failure of that thinking which we characterize here summarily enough as 'metaphysics' (taken from the linguistic usage of Heidegger's later works)? It lies in the fact that . . . metaphysical thinking thinks Being as a constant being-present-at-hand [*Vorhandensein*] and thus cannot do justice to the temporality of factical life's performance." Pöggler, *Martin Heidegger's Path of Thinking*, 29.

5. For Heidegger, the traditional concept of time involves a certain circularity, which Tina Chanter summarizes: "On the one hand, being is understood by reference to certain (and sometimes conflicting) temporal qualities, such as endurance, or immediacy. . . . On the other hand, in asking about the nature, existence, or being of time, philosophers have assumed a certain interpretation of being, restricting in advance the type of being time could have to the kind of being objects have. Time then, the tradition has assumed, is a being, but the very notion of 'a being' is already permeated by temporal assumptions. . . ." Chanter, *Time, Death, and the Feminine*, 26.

6. Stambaugh translates *vulgäre* in different contexts as "ordinary," "common," or "vulgar." The pejorative connotation of "vulgar" makes this a potentially

misleading choice, especially because Heidegger argues in §81 that this understanding of time is legitimate in the context of our concern with things we encounter in the world.

7. As Dastur explains, for Heidegger, "ordinary is not opposed here to philosophical, but characterizes the *public, open* aspect of time." Dastur, *Heidegger and the Question of Time*, xxx. This finally becomes clear in §81 of *Being and Time*, where Heidegger explains how Aristotle's definition of time "makes time thematic in the way that it shows itself in circumspect taking care" (BT 400–1/421). For Heidegger, the commonsense and philosophical concepts of time are both modifications of originary temporality. Thus, while Heidegger distinguishes between the ordinary and the traditional concepts of time, that distinction carries far less importance than the one between time (in both senses) and temporality.

8. Although part two of *Being and Time* was never published, much of the work described in §6 of *Being and Time* as belonging to the destruction of the history of ontology is actually carried out in the 1927 summer course *The Basic Problems of Phenomenology* (GA 24) and the 1927–8 course on Kant's *Critique of Pure Reason* (GA 25), revised and published in 1929 as *Kant and the Problem of Metaphysics* (GA 3). See Friedrich-Wilhelm Von Herrman, "*Being and Time* and *The Basic Problems of Phenomenology*," trans. Parvis Emad and Kenneth Maly, in *Reading Heidegger: Commemorations*, ed. John Sallis (Bloomington: Indiana University Press, 1993), 118–35.

9. On Heidegger's reading of Hegel in §82, see Karin De Boer, *Thinking in the Light of Time*, 255–62.

10. See LQT 206–8/249–51 and 221–3/266–8.

11. According to Heidegger, "Scheler was instrumental in having Bergson translated into German. This recognition of Bergson also brought, within Husserl's work, the *investigations of internal time consciousness*, which are in part published in his later works" (HCT 92/126).

12. These remarks appear to point to Husserl's 1904–5 lectures, which begin with "The Suspension of Objective Time." There, Husserl writes, "From the perspective of objectivity, every experience, just as every real being and moment of being, may have its place in the single objective time—and thus too the experience of the perception and representation of time itself." From a phenomenological standpoint, however, "the time we assume is the *immanent time* of the flow of consciousness, not the time of the experienced world." Husserl, *On the Phenomenology of the Consciousness of Internal Time*, 4–5.

13. The brief discussion of Dilthey, Husserl, and Scheler in §10 of *Being and Time* summarizes some of the points made in §10 and §13 of *History of the Concept of Time*. See HCT 23/28, 92/126, and 126/174.

14. In a discussion near the end of division one, Heidegger admits that some of the ontological characteristics he attributes to Dasein may appear in other living things, adding: "The basic ontological constitution of 'life,' however, is a problem in its own right, and can be developed only reductively and privatively in terms of the ontology of Dasein" (BT 187/194). Heidegger returns to the problem of life in 1929–30. See Martin Heidegger, *The Fundamental Concepts of Metaphysics:*

World, Finitude, Solitude, trans. William McNeill and Nicholas Walker (Bloomington: Indiana University Press, 1995).

15. The last three chapters of *Being and Time*, division two, deal with temporality and "everydayness" (*Alltäglichkeit*), temporality and "historicity" (*Geschicktlichkeit*), and temporality and "within-timeness" (*Innerzeitigkeit*). The final chapter is where Heidegger offers his account of how time, as we ordinarily understand it, is related to temporality. It is also where he offers an alternative to Bergson's interpretation of time.

16. See also HCT 5/7: "both historical reality and natural reality are continuities that run their course *in time* and are traditionally understood as such. . . . Viewed simply from the outside, history and nature are temporal. . . . In a very schematic and crude way, time already announces itself as *one* 'index' for the differentiation and delimitation of being as such."

17. See especially §§79–81, BT 387/406ff, which I discuss below.

18. As in his discussion of being-in-the-world, Heidegger is concerned both to distinguish the constitutive parts of the structure and to account for their unity. He thus uses the familiar term "care" as a *structural condensation* of items already treated separately. See Wood, *The Deconstruction of Time*, 173.

19. As John Sallis explains, there is a double bind in making assertions about temporality: "it *must be said* that temporality *is* . . . and yet, it *cannot be said* that it *is*, since it is not a being at all." Sallis, *Echoes*, 59. To emphasize this, Heidegger uses the expression, "temporality *temporalizes*."

20. The centrality of death to Heidegger's interpretation of temporality in *Being and Time* sets him apart from Bergson and most other philosophers. Emmanuel Levinas recognized both the extent to which Heidegger followed Bergson in distinguishing between time and temporality, and that to which he parted ways with Bergson in understanding authentic temporality as a way of being-towards-death. In the lecture "Inside Heidegger: Bergson," Levinas compares their approaches to time:

> [For Heidegger] It is in exposing oneself to one's most proper possibility, in the power to die, that meaning is projected—the same meaning toward which every project is aimed: the future, or what is to-come [*l'àvenir*]. In this way, without recourse to quantitative notions of time, the future and an originary notion of time are sketched out, and one more properly time than was everyday time.
>
> As in Bergson, the idea that there are various levels of time is affirmed here. The entire Western tradition approaches time through measurement (time is the number of movement, says Aristotle). For Bergson, linear time is a spatialization of time in view of acting upon matter, which is the work of intelligence. Originary time he calls *duration*; this is a becoming which each instant is heavy with all of the past and pregnant with the whole future. Duration is experienced by a descent into self. Each instant is there; nothing is definitive since each instant remakes the past.

> For Heidegger, originary time, or the time of the being-there that is fulfilled in the human, describes the finitude of being-there [*Da-sein*]. It is fulfilled in anxiety and dispersed in the everyday. For Heidegger, infinite time is deduced from original finitude. For Bergson, finitude and unsurpassable death are not inscribed in duration. Death is inscribed in the degradation of energy. . . . life is duration, the vital impulse [*élan vital*], and these must be thought together with creative freedom. (Emmanuel Levinas, *God, Death, and Time*, trans. Bettina Bergo [Stanford: Stanford University Press, 2000], 55)

It is surprising that Heidegger does not take the opportunity in *Being and Time* to take issue with Bergson for leaving death out of his account of duration. On Heidegger's analysis of death in relation to his conception of time, see also Dastur, *Heidegger and the Question of Time*, 17–51; Wood, *The Deconstruction of Time*, 179–220; Chanter, *Time, Death, and the Feminine*, 123–39; and White, *Time and Death*, 53–91.

21. According to Françoise Dastur, "the German term that Heidegger uses, *Zeitigung*, expresses 'the work' of time, *Zeit*, and literally means 'maturation.'" Dastur, *Heidegger and the Question of Time*, 35. In the conclusion I show how Heidegger clarifies this idea with the help of a Bergsonian image.

22. The German *Zukunft*, like the French *avenir*, literally means "to come."

23. Insofar as what we plan to do and who we hope to be give meaning and value to what we have done and who we have been, "What matters about the past depends on what matters about the future." White, *Time and Death*, 99.

24. As Dastur explains, "because Dasein is not an entity that is present-at-hand or given beforehand (*vorhandenes*), it is never 'past' in the strict sense." Dastur, *Heidegger and the Question of Time*, 36.

25. The equiprimordiality of the ecstasies of temporality means that, as Tina Chanter puts it, "Heidegger interprets the past and future not as derivative versions of the present—not as a past that once was present, nor as a future that is yet to become present. Rather, each ecstasis has its own dynamic, and each ecstasis is also a dimension of each of the other ecstases." Chanter, *Time, Death, and the Feminine*, 126.

26. On this point, David Couzens Hoy sees a clear divergence: "Where Heidegger differs most significantly from Bergson is in (1) the priority that Heidegger gives to the future in contrast with Bergson, for whom the future is not real, and (2) the account that Heidegger gives of historicity and temporality." David Couzens Hoy, *The Time of Our Lives: A Critical History of Temporality* (Cambridge: MIT Press, 2009), 197. While it is true that Bergson gives more attention to the past as such in *Matter and Memory*, it is also true that he consistently emphasizes the openness and unforeseeability of the future. Levinas goes so far as to suggest that with Bergson's conception of duration as "the uninterrupted upsurge of novelty" (CM 18/1259), we find that "the thematization of the future 'being-of-the-being' appears in Bergsonism as in *Being and Time*." Emmanuel Levinas, "The Old and the New," in *Time and the Other*, trans. Richard A. Cohen (Pittsburgh: Duquesne

University Press, 1987), 130n. It is not so much that the future is unreal, for Bergson, as that it is irreducible to the present.

27. In *The Basic Problems of Phenomenology* (SS 1927), Heidegger points out that Aristotle leaves this question open. See BPPb 254/359.

28. Heidegger signals the connection between this characteristic of world time and originary temporality etymologically: the "publicness" (*Öffentlichkeit*) of average, everyday being-with-others and the "making public" (*Veröffentlichung*) of time in circumspect taking care are grounded in temporality, which is "ecstatically open" (*ekstatisch offen*).

29. This argument about world time parallels one that appears in §69c about the world itself. Heidegger writes: "If the 'subject' is conceived ontologically as existing Dasein, whose being is grounded in temporality, we must say then that the world is 'subjective.' But this 'subjective' world, as one that is temporally transcendent, is then 'more objective' than any possible 'object' " (BT 349/366).

30. Heidegger pursues the connection between temporality and the *a priori* in *The Basic Problems of Phenomenology* (SS 1927). See BPPb 324/461ff.

31. William Blattner argues that Heidegger, perhaps under Bergson's influence, remains an "explanatory, transcendental idealist about time" (Blattner, *Heidegger's Temporal Idealism*, 276). This reading is difficult to reconcile with Heidegger's insistence that world time and, *a fortiori*, originary temporality is prior to the distinction between subject and object. On the extent to which Heidegger resists Kant's transcendental idealism, see François Raffoul, *Heidegger and the Subject*, trans. David Pettigrew and Gregory Recco (Atlantic Highlands, NJ: Humanities Press, 1998), 70ff.

32. As Heidegger argues in §6 of *Being and Time*, the obviousness or self-evidence of traditional concepts often conceals the originary experiences from which they were drawn. With respect to time, this is the case with both the kind of experience Aristotle privileges—the experience of motion—and the ontological orientation guiding his analysis.

33. This argument appears in greater detail in §19 of *The Basic Problems of Phenomenology* (SS 1927), where Heidegger writes: "Aristotle understands being in the sense of extantness [*Vorhandenheit*]. If you take being in this sense, then you have to say that the now which is no longer extant in the sense of the bygone now and the now which is not yet extant in the sense of the now yet to come, are not—that is, are not extant. Seen in this way, what *is* in time is only the now that is extant in each now. Aristotle's aporia with reference to the being of time—which is still the principal difficulty today—derives from the concept of being as equal to being extant" (BPPb 272/386).

34. Heidegger imagines Aristotle using a pointer to demonstrate how time shows up in connection with a moving thing. Here, it sounds as if he could also be referring to the minute hand of a clock.

35. See also MFL 138/174: "The phenomenon of Dasein's dissemination in space is seen, for example, in the fact that all languages are shaped primarily by spatial meanings."

36. As John Sallis has shown, the clear exception is Heidegger's description of temporality qua ecstatic as the "primordial outside-of-itself." See Sallis, *Echoes*, 60. My claim is not that Heidegger *succeeds* in thinking non-spatially about temporality, but that he makes a sustained effort to do so.

37. William Blattner acknowledges, "Bergson's official doctrine . . . that time consists of qualitative, nonsuccessive tenses . . . we may view as having influenced Heidegger deeply" (*Heidegger's Temporal Idealism*, 269). He also provocatively suggests that "Heidegger's achievement was to have taken Bergson's one enduring insight—that nonsuccessive time would have to be the tenses shorn of their successive content—and to have developed it into an elaborate theory" (270). However, with respect to Bergson's distinction between "pure duration" and "spatialized time," Blattner stresses: "This could not be farther [*sic*] from what Heidegger is up to" (269n). In my view, while Heidegger argues, against Bergson, that space is not the issue, he also engages in a thorough de-spatialization of temporality.

38. In *The Basic Problems of Phenomenology* (SS 1927), he uses the term "dimension" to refer to "a general notion of *stretch*" that is implicit in Aristotle's account of the relation between time and motion. See BPPb 242/343 and Chapter 4.

39. The view that temporality is ontologically prior to spatiality is one that Heidegger reconsiders after *Being and Time*. In the 1930s he begins using the term "Time-Space" (*Zeit-Raum*) to designate the origin of both time and space. For example, in the *Beiträge* he writes: "Space and time, each represented for itself and in the usual connection, themselves arise from time-space, which is more originary than they themselves and their calculatively represented connection." Martin Heidegger, *Contributions to Philosophy (From Enowning)*, trans. Parvis Emad and Kenneth Maly (Bloomington: Indiana University Press, 1999), 259. He reiterates this shift in his late, retrospective lecture "Time and Being" (1962), where he says: "Time-space is now the name for the openness which opens up in the mutual self-extending of futural approach, past and present." Martin Heidegger, *On Time and Being*, trans. Joan Stambaugh (Chicago: University of Chicago Press, 1972), 14. On the development of his thinking about the relationship between time and space, see Dastur, *Heidegger and the Question of Time*, 63–69; Sallis, *Echoes*, 69–75; and Miguel de Beistegui, *Truth and Genesis: Philosophy as Differential Ontology* (Bloomington: Indiana University Press, 2004), 141–66, and *The New Heidegger* (London: Continuum, 2005), 60–96.

40. A notable exception: Hannah Arendt, in a discussion of Heidegger's analysis of the self in *Being and Time*, notes its similarities with that of Bergson, who "had posited, only a few decades before Heidegger, the co-existence of two selves, the one social (Heidegger's 'They') and the other 'fundamental' (Heidegger's 'authentic')." Hannah Arendt, *The Life of the Mind*, vol. 2. (New York: Harcourt, Brace, & Company, 1978), 183.

41. Heidegger's analysis of the concept of reality in §43 of *Being and Time* reveals another possible line of criticism of Bergson, who commonly speaks of "real duration," "real movement," and "the real" (as opposed to "the possible"). See BT 193–204/200–12.

Chapter 4

1. After abandoning his plan to publish the second part of *Being and Time*, Heidegger omits this line. My citation refers to the German text only, as this line does not appear in Stambaugh's translation of *Being and Time*.

2. For a detailed account of the extent to which Basic Problems follows the plan of *Being and Time*, see Von Herrman, "*Being and Time* and The Basic Problems of Phenomenology."

3. On Heidegger's reading of Aristotle in *Basic Problems*, see Tina Chanter, "Heidegger's Understanding of the Aristotelian Concept of Time," in *Interrogating the Tradition: Hermeneutics and the History of Philosophy*, ed. Charles E. Scott and John Sallis (Albany: State University of New York Press, 2000), 131–57, and "Heidegger's Critique of Metaphysical Presence," in *Time, Death, and the Feminine*, 123–39; John Ellis, "Heidegger, Aristotle, and Time in *Basic Problems* 19," in *Interrogating the Tradition*, 159–78; Stephen Crocker, "The Oscillating Now: Heidegger on the Failure of Bergsonism; and Robin Durie, "Splitting Time: Bergson's Philosophical Legacy."

4. Heidegger lists the date of publication for Bergson's *Essai* (*Time and Free Will*) incorrectly as 1888. Although Bergson signed his foreword "February 1888," it was not actually published until 1889. See Bergson, *Œuvres*, 1485.

5. Heidegger also mentions Bergson's continuation of his efforts to think in terms of duration in *Creative Evolution* (1907) and his response to Einstein's theory of relativity in *Duration and Simultaneity* (1922). On the way that the debate between Einstein and Bergson figures into *Being and Time*, see David Scott, "The 'Concept of Time' and the 'Being of the Clock.' " It is particularly interesting that Heidegger does not cite Bergson's *Matter and Memory* (1896), which he had mentioned in his 1925–6 *Logic* course as the location of "the essential and enduring element of his philosophical work" (LQT 207/250). I take a closer look at this omission in Chapter 5.

6. In the discussion of the history of ontology in §6 of *Being and Time*, Heidegger says this about the ontological tradition: "What has been handed down is handed over to obviousness; it bars access to those original 'wellsprings' out of which traditional categories and concepts were in part genuinely drawn. The tradition even makes us forget such a provenance altogether" (BT 20–21/21). This means that we have to turn our attention not only from beings to being, but also from our concepts to those who defined them and the experiences in relation to which they were defined. Thus, we cannot succeed in rethinking time, for Heidegger, without a destruction—or better, deconstruction—of the traditional concept of time as it appears in Aristotle.

7. As Robert Bernasconi emphasizes, there are two senses of "tradition" in Heidegger: On the one hand, "[t]radition as a weight that drags one back into the past and from which one must somehow shake oneself loose, he called *Tradition*. Heidegger is clear that this conception arises only from what he calls an inauthentic conception of temporality. On the other hand, tradition as a heritage in terms of which the possibilities of authentic existing are disclosed and taken over (BT

435/383). This is the task of the famous *Destruktion* or destructuring of the history of philosophy. . . . Destructuring does not attempt to transcend these limits imposed by the tradition, but it does attempt to mark those limits in such a way as to stake out its positive possibilities" (BT 22/44). Robert Bernasconi, "Heidegger and the Invention of the Tradition," *Journal of the British Society for Phenomenology* 26, no. 3 (October 1995): 242. See also Robert Bernasconi, "Repetition and Tradition: Heidegger's Destructuring of the Distinction between Essence and Existence in *Basic Problems of Phenomenology*," in *Reading Heidegger from the Start*; and Robert Bernasconi, "Heidegger's Destruction of *Phronesis*," *The Southern Journal of Philosophy* 28 (1990): 127–47.

8. In the interpretation of Aristotle offered in *Basic Problems*, Heidegger makes it clearer than he does in *Being and Time* that "Aristotle exceeds a tradition which he also represents" (Bernasconi, "Heidegger's Destruction of *Phronesis*," 139).

9. As De Boer explains, "Heidegger's reading of Aristotle is in fact characterized by a double deconstruction: first, the meaning of Aristotle's thinking is 'freed' from prevailing, concealing interpretations; second, the meaning that emerges in this way is itself revealed in its confinement." De Boer, *Thinking in the Light of Time*, 24. Not only does Heidegger defend Aristotle against faulty interpretations like the ones he attributes to Bergson, but against the notion that recent philosophers have gone well beyond his thinking, Heidegger also tries to show that, as Chanter puts it, "Aristotle has already caught sight of the horizonal structure of temporality." Chanter, *Time, Death, and the Feminine*, 129.

10. *Physics*, 217^{b}31–32.

11. *Physics*, 220^{a}25–26.

12. *Physics*, 223^{a}22–28.

13. In their translations of *Being and Time*, Stambaugh renders *vorhanden* as "objectively present," while Macquarrie and Robinson use the more literal "present-at-hand." In *Basic Problems*, Hofstadter opts for "extant" in accordance with Heidegger's use of *Vorhandensein* and *Vorhandenheit* to translate the Latin *existentia*. See BPPb 28/36. To maintain consistency between Chapter 3 and the discussion that follows, I have changed "extant" to "objectively present" whenever it is used to translate *vorhanden*.

14. Recall Heidegger's characterization of Greek ontology in §6 of *Being and Time*: "Here it becomes evident that the ancient interpretation of the being of beings is oriented toward the 'world' or 'nature' in the broadest sense and that it indeed gains its understanding of being from 'time.' The outward evidence of this—but of course *only* outward—is the determination of the meaning of being as παρουσία [*parousia*] or ουσία [*ousia*], which ontologically and temporally means 'presence' ['*Anwesenheit*']. Beings are grasped in their being as 'presence'; that is to say, they are understood with regard to a definite mode of time, the '*present*' ['*Gegenwart*']" (BT 24/25). As a result of this understanding of being, "which is oriented toward time in an inexplicit and naïve way," ancient Greek philosophers regard time as "one being among others" without being aware of "the fundamental ontological function of time" (BT 25/26). As I have shown, Heidegger goes on to argue in §81 that this orientation becomes especially clear in Aristotle when

he defines time on the basis of the ordinary understanding of it as "a succession of constantly 'present' ["*vorhanden*"] nows that pass away and arrive at the same time" (BT 401/422).

15. Chanter formulates the paradox clearly: "Metaphysics has obscured from itself the fact that it has used time to think being, and being to think time, without acknowledging that the very interpretations of both time and being that are in operation feed off one another in subterranean ways." Chanter, *Time, Death, and the Feminine*, 2; see also 25–27.

16. Robin Durie argues, against Heidegger's critique of Bergson and with the help of Deleuze's reading, that "the opposition between limit and transition, point and continuum, constitutes what Bergson would in fact call a 'false problem.'" Durie, "Splitting Time," 164. Bergson replaces Aristotle's question of how the now can be both a limit and a transition between the past and the future with the question of how time (that is, duration) can be both heterogeneous and continuous.

17. Stephen Crocker claims that in *Basic Problems*, "the phenomenological exposition of both time and motion is developed in direct opposition to the thesis which Heidegger attributes to Bergson, namely that Aristotle derives time from space." Crocker, "The Oscillating Now," 411. If this is true, then Heidegger's reading of Aristotle in these lectures is shaped by his reading of Bergson: before he highlights the ontological presuppositions of Aristotle's treatment of time, Heidegger wants to show why it is wrong to interpret Aristotle as reducing time to space.

18. Crocker discusses how more recent commentators on Aristotle, specifically Edward Hussey, G. E. L. Owen, and David Ross, read him as performing a "derivation of time from space." Crocker, "The Oscillating Now," 410. Although this is a standard interpretation, the only interpreter of Aristotle whom Heidegger mentions by name in §19 of *Basic Problems* is Bergson.

19. See, for example, *Physics*, 219^{a}4–6: "if any movement takes place in the mind we at once suppose that some time has indeed elapsed; and not only that but also, when some time is thought to have passed, some movement also along with it seems to have taken place." Heidegger explains later that, in Aristotle's view, "[t]he mind, too, has the character of a moving thing. Even when we are not experiencing something moving in the sense of some entity presently at hand, nevertheless motion taken in the broadest sense, hence time, is unveiled for us in experiencing our own self" (BPPb 358/254).

20. As I discussed in Chapter 3, this structure makes a brief appearance in §79 of *Being and Time* in connection with one of the four characteristics of world time, "spannedness." Interestingly, this is the only point in Heidegger's interpretation of temporality where he discusses duration: "The making present that awaits and retains interprets a 'during' with a 'span,' only because in so doing it is disclosed to *itself* as being ecstatically *stretched along*" (BT 390/409).

21. The term "dimension" in *Basic Problems* designates a "stretching out" prior to any distinction between spatial extension and temporal duration in much the same way that "existence" formally indicates the being of *Dasein* as "standing

out" in a certain way. On Heidegger's use of formal indication in his early works, see Kisiel, *The Genesis of Heidegger's* Being and Time, 50–55.

22. *Physics*, 219^{a}11.

23. According to John Ellis, "*akolouthein* is the key to Heidegger's interpretation of Aristotle's time treatise. He understands it in such a way that not only links the stretching out of motion to magnitude and time, but also points in the direction of original temporality." Ellis, "Heidegger, Aristotle, and Time," 173–4.

24. *Physics*, 219^{a}15.

25. Heidegger's interpretation thus returns to a crucial point stressed in §81 of *Being and Time*, namely that Aristotle's definition of time conceptualizes the way time is ordinarily understood "as a succession of constantly 'present' ['*vorhanden*'] nows that pass away and arrive at the same time . . . as a sequence, the 'flux' of 'nows,' as the 'course of time' " (BT 401/422). There, Heidegger argues that in this concept, nows are stripped of their relations with one another and the phenomenon of the now is "cut off in its complete structure" (BT 405/427). But in *Basic Problems*, Heidegger attends much more closely to the peculiar structure of the now described by Aristotle in *Physics* IV, 11, 219^{b}10ff.

26. *Physics*, 220^{a}5.

27. Crocker interprets Heidegger somewhat differently as claiming that the now sometimes functions as a boundary between past and future, and sometimes as a transition: "With the aid of 'the now' we are able to mark the limits where one motion ends and another begins. But because time is intrinsically transitory and always contains a reference to what is 'not yet' and what is 'no longer,' the now is always beyond the limits of the motion it marks. We might say that the now 'oscillates' in and out of periodic and transitory functions." Crocker, "The Oscillating Now," 406. See also 413–15 and 418. As I read Heidegger, his claim is that what is peculiar or distinctive about the now is the structure that enables it to function in *both* ways. Because that structure is ontologically prior to the representation of nows as points in succession, though, it makes little difference whether we regard the now as oscillating back and forth between these functions or serving both functions at once.

28. Stambaugh renders *Innerzeitigkeit* as "within-timeness," while Hofstadter translates it as "intratemporality," reflecting the contrast with *Ausserzeitigkeit* or "extratemporality": "So far as motion or rest can be measured by time, and to be measured by time means "to be in time," the moving or resting thing, and only it, is in time. For this reason we may say that geometrical relationships and their contents are extratemporal, because they are not in motion and consequently also are not at rest" (BPPb 253/357). See also BT 18/18 and 398–9/419–20.

29. As Ellis argues, "Heidegger twice *reverses* fundamental priorities of Aristotle's, once in the way that he understands ἀκολουθεῖν (*akolouthein*) to be at work in the series, magnitude-motion-time, and again in his argument that the primary sense of the 'now' is dimensional" (Ellis, "Heidegger, Aristotle, and Time," 160). Ellis laments that "the difficulties involved in attributing these reversals to Aristotle

are insurmountable" (161), but he rightly focuses on Heidegger's reason for reading Aristotle against the grain, namely to show the extent to which Aristotle already recognizes features of originary temporality and world time.

30. "The distinction between before and after holds primarily, then, in place." *Physics*, 219^{a}15.

31. *Physics*, 220^{b}21.

32. *Physics*, 233^{b}32–234^{a}23. This reversal is less obvious because "Heidegger does not indicate the unusual nature of his interpretation, nor does he discuss the central passage relevant to the issue, namely *Physics* 6.3." Ellis, "Heidegger, Aristotle, and Time," 173. Crocker is too generous in accepting that "while the now may appear as segmented and point-like, Aristotle indicates that it is in fact defined by transition." Crocker, "The Oscillating Now," 413. I share Ellis's view that Heidegger reverses Aristotle's position and ignores the discussion of the now in *Physics* VI, 3.

33. *Physics*, 221^{a}27–29.

Chapter 5

1. In *The Metaphysical Foundations of Logic*, Heidegger elaborates: "Recently Bergson tried to conceive the concept of time more originally. He made it more clear than any previous philosopher that time is interwoven with consciousness. . . . He developed his interpretation of time on the basis of the traditional concept of consciousness, of Descartes' *res cogitans*" (MFL 149/189). Later in the same course, Heidegger also criticizes Husserl for interpreting time as "something immanent . . . something internal, 'in the subject.' Hence the title, 'internal time-consciousness' " (MFL 203–4/263–4). I return to these passages in the Conclusion.

2. See also LQT 204–5/246.

3. In the introduction to *Matter and Memory*, Bergson argues that "the classical problem of the relations of soul and body" can be solved with the help of research into memory, which is "the intersection of mind and matter" (MM 13/164). Instead of emphasizing the difference between scientific and philosophical inquiry, he proposes: "Without denying to psychology any more than to metaphysics, the right to make itself into an independent science, we believe that each of these two sciences should set problems to the other and can, in a measure, help it to solve them. How should it be otherwise, if psychology has for its object the study of the human mind working for practical utility, and if metaphysics is but this same mind striving to transcend the conditions of useful action and come back to itself as to a pure creative energy?" (MM 15/167). This appears to stand in sharp contrast to Heidegger's insistence in *Being and Time*, §10, that fundamental ontology be kept apart from the "ontic sciences" of anthropology, psychology, and biology. Heidegger's aim is to rethink the ontological foundations of these sciences, and Bergson draws from research in psychology for the very purpose of attacking common assumptions about its basic concepts. Bergson's concern with ontological issues becomes clearer as

he moves from psychological hypotheses about the nature of memory to arguments about the existence or nonexistence of the past.

4. My interpretation of *Matter and Memory* builds on the work of many others, including Leonard Lawlor, *Early Twentieth-Century Continental Philosophy* (Bloomington: Indiana University Press, 2012); *The Challenge of Bergsonism: Phenomenology, Ontology, Ethics* (London: Continuum, 2003); *Thinking through French Philosophy: The Being of the Question* (Bloomington: Indiana University Press, 2003); and "The Ontology of Memory: Bergson's Reversal of Platonism," *Epoché* 8 (2003): 69–102; Alia Al-Saji, " 'A Past Which Has Never Been Present': Bergsonian Dimensions in Merleau-Ponty's Theory of the Prepersonal, *Research in Phenomenology* 38 (2008): 41–71; "The Memory of Another Past: Bergson, Deleuze, and a New Theory of Time," *Continental Philosophy Review* 37 (2004): 203–39; and "A Phenomenology of Critical-Ethical Vision: Merleau-Ponty, Bergson, and the Question of Seeing Differently," *Chiasmi International* 11 (2009): 375–98; Valentine Moulard-Leonard, *Bergson-Deleuze Encounters: Transcendental Experience and the Thought of the Virtual* (Albany: State University of New York Press, 2008); Michael R. Kelly, "Husserl, Deleuzean Bergsonism and the Sense of the Past in General," *Husserl Studies* 24 (2008): 15–30; Suzanne Guerlac, *Thinking in Time: An Introduction to Henri Bergson* (Ithaca: Cornell University Press, 2006); Mark S. Muldoon, *Tricks of Time: Bergson, Merleau Ponty and Ricoeur in Search of Time, Self and Meaning* (Pittsburgh: Duquesne University Press, 2006); Rudolf Bernet, "A Present Folded Back on the Past (Bergson)," *Research in Phenomenology* 15 (2005): 55–76; Elizabeth Grosz, *Time Travels: Feminism, Nature, Power* (Durham: Duke University Press, 2005) and *The Nick of Time: Politics, Evolution, and the Untimely* (Durham: Duke University Press, 2004); Stephen Crocker, "The Past Is to Time What the Idea Is to Thought or, What Is General in the Past in General?," *Journal of the British Society for Phenomenology* 35 (2004): 42–53; Keith Ansell Pearson, *Philosophy and the Adventure of the Virtual: Bergson and the Time of Life*; Keith Ansell Pearson and John Mullarkey, introduction to *Henri Bergson: Key Writings* (New York: Continuum, 2002), 1–45; Steven G. Smith, "Mind-Matter Inversions: Bergson's Conception of Mental and Material Actuality," *Southern Journal of Philosophy* 40 (2002): 295–314; Leszek Kolakowski, *Bergson* (South Bend, IN: St. Augustine's Press, 2001); Robin Durie, "Splitting Time: Bergson's Philosophical Legacy," *Philosophy Today* 44 (2000): 152–68; John Mullarkey, *Bergson and Philosophy*; Frédéric Worms, "*Matter and Memory* on Mind and Body: Final Statements and New Perspectives," trans. Pelagia Goulimari, in *The New Bergson*, ed. John Mullarkey (Manchester: Manchester University Press, 1999), 88–98; and Marie Cariou, "Bergson: The Keyboards of Forgetting," trans. Melissa McMahon, in *The New Bergson*; Frédéric Worms, *Introduction à* Matière et mémoire *de Bergson: suivie d'une brève introduction aux autres livres de Bergson* (Paris: Presses Universitaires de France, 1997); F. C. T. Moore, *Bergson: Thinking Backwards* (Cambridge: Cambridge University Press, 1996); Gilles Deleuze, *Bergsonism*, trans. Hugh Tomlinson and Barbara Habberjam (New York: Zone Books, 1991) and "Bergson's Conception of Difference," in *Desert Islands and Other Texts*

(1953–1974), trans. Mike Taormina (Los Angeles: Semiotext(e), 2003), 32–51; and Jean Hyppolite, "Various Aspects of Memory in Bergson," trans. Athena V. Colman, in Leonard Lawlor, *The Challenge of Bergsonism*, 112–27.

5. It is common, though controversial, to distinguish between "early" and "late" Heidegger, based on the self-described "turn" (*Kehre*) in his thinking. This is not the case with respect to Bergson. However, it has been argued that Jean Hyppolite does make such a distinction and that Gilles Deleuze follows suit. According to Crocker, "Bergson's 'turn' as we might call it, begins with his new investigations of memory." Crocker, "The Past Is to Time What the Idea Is to Thought," 44. Hyppolite defends Bergson from some of the main criticisms of *Time and Free Will* by emphasizing the development of his thinking in *Matter and Memory*, one aspect of which Deleuze characterizes in this way: "Duration seemed to him to be less and less reducible to a psychological experience and became instead the variable essence of things, providing the theme of a complex ontology." Deleuze, *Bergsonism*, 34. Deleuze argues that in *Matter and Memory* Bergson comes to think of "psychological duration" as "only a clearly determined case, an opening onto ontological duration" (48–49), and psychology becomes "an opening onto ontology, a springboard for an 'installation' in Being" (76). For more on this aspect of Deleuze's interpretation, see Ansell Pearson, *Philosophy and the Adventure of the Virtual*, 35–38. Merleau-Ponty notes this development in Bergson's thinking as well: "Duration is not simply change, becoming, mobility; it is being in the vital, active sense of the term. Time is not put in place of being; it is understood as being coming to be, and now it is the whole of being which must be approached from the side of time. This was evident when *Matter and Memory* appeared, or at least it should have been. . . . it is in this book that the field of duration and the practice of intuition are enlarged in a decisive way." Merleau-Ponty, *Signs*, 184. When I refer to an "ontological turn" in *Matter and Memory*, my intention is not to distinguish between two phases of Bergson's career, but rather to emphasize how his thinking develops around questions about what it means for something to exist and why we deny existence to the past.

6. In the analysis that follows, I focus especially on what Deleuze calls the "temporalizing" aspect of Bergson's method. See Deleuze, *Bergsonism*, 31–35.

7. In *The Challenge of Bergsonism*, Leonard Lawlor attempts "to take seriously the standard Heidegger has set for ontology and try to see whether Bergson's metaphysics of memory lives up to it" (28). According to Lawlor, when Heidegger claims that Bergson reverses Aristotle's concept of time, "Heidegger's use of the word 'reversal' here anticipates . . . [his claim] that Nietzsche merely 'reverses' Platonism without 'twisting free' of it" (28). With this in mind, we can ask, "Is it the case that Bergson *merely* reverses Platonism without twisting free of it?" (28). Lawlor argues that Bergsonism is not a "new subjectivism," as Heidegger claims that all modern metaphysics is, for he "twists free" of Platonism with his image of the cone of memory. Expanding on Lawlor's interpretation, I aim to show how Bergson's efforts to rethink the relationship between mind and body in terms of time—or

duration—lead him to reject certain assumptions about being and consciousness that Heidegger also criticizes.

8. Paul and Palmer translate *l'esprit* as "spirit" in most cases, but I use "mind" to emphasize the connection that Bergson attempts to maintain between metaphysics and psychology. When quoting from their translation, though, I have not changed "spirit" to "mind."

9. Deleuze offers the first chapter of *Matter and Memory* as a paradigm of Bergsonian intuition as a "method of division," which demonstrates how perception and memory differ in kind. Rather than treating them as two mental faculties, Bergson divides human experience into an objective line leading from perception to matter and a subjective one leading from memory to mind. See Deleuze, *Bergsonism*, 24–26.

10. When Bergson speaks of "pure" phenomena—duration, perception, memory—he means tendencies that run counter to one another in principle but are mixed together in experience. As Deleuze writes, "Only that which differs in kind can be said to be pure, but only *tendencies* differ in kind. . . . This is the Bergsonian leitmotif: People have seen only differences in degree where there are differences in kind." *Bergsonism*, 22–23. What needs to be thought, according to Bergson, are differences in kind *and* the exchange between them, for instance, the "endosmosis" between consciousness and the world described in *Time and Free Will*.

11. On Bergson's conception of the image and his reasons for beginning *Matter and Memory* with the hypothesis that reality consists of images, see Lawlor, *The Challenge of Bergsonism*, 1–26.

12. Jean Hyppolite argues that Bergson's focus on the body indicates his adoption of a different standpoint from that of *Time and Free Will*: "In its widest scope, *Matter and Memory* raises the question of *incarnation*. . . . we are no longer considering merely pure and undivided duration, but the relationship of this duration to things. . . . This is why *Matter and Memory* starts from the world, then the body, and not from interior duration." Hyppolite, "Various Aspects of Memory in Bergson," 115.

13. In this way, conscious perception is, for Bergson, a kind of anticipation. As he later writes in *Creative Evolution*, "a choice involves the anticipatory idea of several possible actions. Possibilities of action must therefore be marked out for the living being before the action itself. Visual perception is nothing else: the visible outlines of bodies are the design of our eventual action on them" (CE 96/577).

14. Deleuze views this as a crucial difference between Bergson and phenomenology. See Deleuze, *Cinema I*, 60–61.

15. In this respect, at least, Bergson's theory of pure perception anticipates Heidegger's account of being-in-the-world, particularly his claim that Dasein does not need to escape from a subjective sphere in order to grasp objects because it is always already in an environment, together with beings other than itself.

16. Alia Al-Saji uses this aspect of Bergson's theory of perception as a point of departure for a critical examination of "objectifying habits of seeing." See Al-Saji, "A Phenomenology of Critical-Ethical Vision," 385–7.

17. For a typology of memory that draws on both phenomenology and Bergsonism, see Edward S. Casey, *Remembering: A Phenomenological Study*, 2nd ed. (Bloomington: Indiana University Press, 2000), 48–64.

18. In their translation of *Matter and Memory*, Paul and Palmer always translate *mémoire* as "memory," but they translate *souvenir* as "recollection" in some cases and as "memory" in others. *Mémoire* typically refers to the faculty for remembering, while *souvenir* can refer to a past event, a trace of the past, or an act of remembering. Bergson uses *souvenir* for images recalled from the past—"*images-souvenir*"—and for events preserved by memory—"*souvenir pur*." Paul and Palmer use "memory-images" and "pure memory" to translate these terms. Instead of changing these to "recollection-images" and "pure recollection" to signal the use of *souvenir*, I follow their usage unless otherwise indicated.

19. See Lawlor, *The Challenge of Bergsonism*, 32.

20. This differs from Husserl's understanding of retention as the conscious act that keeps the immediate past connected to the present. For Bergson, bodily movements are responsible for an unconscious, automatic prolongation of the past into the present, which, although geared toward action and life, would fit under the heading of what Husserl calls "passive synthesis." Bergson does not distinguish between retention and recollection as Husserl does, but he seems to have retention in mind whenever he describes the prolongation of the past into the present.

21. Of course, Bergson offers several of his own images in *Matter and Memory*, chapter three, most famously the "cone of memory," and these images are helpful as long as we keep in mind that they are symbols for *movement*.

22. Heidegger discovers a similar description by Lotze of "two unequal arms of non-being stretching out in both directions: one into the past and the other into the future," and notes that "Lotze's image characterizes very clearly the way the real accent of temporal being falls on the now, the now-present" (LQT 205/247). For Bergson, however, the present that includes "both a perception of the immediate past and a determination of the immediate future" (MM 138/280) is more real, more "concrete," than the instant; it is "my present."

23. While the contrast between the actual, lived present and the ideal present is familiar from Husserl and William James, we can see how Bergson's view differs from any theory of "time-consciousness" in the role played by the body.

24. Deleuze demonstrates how Bergson's argument challenges the privilege of presence as Heidegger understands it: "We have great difficulty in understanding the survival of the past in itself because we believe that the past is no longer, that it has ceased to be. We have thus confused Being with being-present. . . . The past . . . has ceased to act or to be useful. But it has not ceased to be." Deleuze, *Bergsonism*, 55.

25. In the introduction to *Matter and Memory*, Bergson writes, "There are then, in short, diverse *tones* of mental life, or, in other words, our psychic life may be lived at different heights, now nearer to action, now further removed from it, according to the degree of our *attention to life*" (MM 14/166). Deleuze explains what Bergson means by "attention to life" in temporal terms "the adaptation of the

past to the present, the utilization of the past in terms of the present." Deleuze, *Bergsonism*, 70.

26. In contrast to Heidegger's constant preoccupation with ontology, Bergson apologetically notes that the meaning of existence is "a problem we can only glance at, for otherwise it would lead us step by step into the heart of metaphysics (MM 146–7/288).

27. Heidegger was impressed enough by this description that he used part of it as the epigraph for his 1919–20 winter course *Basic Problems of Phenomenology*. Nevertheless, he does not examine *Matter and Memory* in the course, but treats Bergson as one of several leading philosophers of life, including Dilthey, James, and Simmel. See Chapter 1 for a discussion of Heidegger's remarks about Bergson in this course.

28. On Deleuze's interpretation, Bergson is attempting to break the habit of thinking of the relationship between the past and the present as one of succession. For Bergson, he writes, "The past and the present do not denote two successive moments, but two elements which coexist. . . . The past does not follow the present, but on the contrary, is presupposed by it as the pure condition without which it would not pass." Consequently, in contrast to Heidegger's interpretation, "Bergsonian duration is, in the final analysis, defined less by succession than by coexistence." Deleuze, *Bergsonism*, 59–60.

29. Merleau-Ponty is best known for this expression, some version of which also appears in Deleuze, Derrida, and Foucault. See Lawlor, *The Challenge of Bergsonism*, 54, and *Thinking through French Philosophy*, 88–92; see also Alia Al-Saji, "A Past Which Has Never Been Present.

30. As Lawlor explains, "the cone is supposed to be an image of movement. The cone really symbolizes a dynamic process." Lawlor, *The Challenge of Bergsonism*, 47.

31. Bergson stresses that pure memory and pure action, like pure perception, are not actual experiences: "these are only two extreme limits . . . which are never really reached in practice. There is not, in man at least, a purely sensori-motor state, any more than there is in him an imaginative life without some slight activity beneath it. Our psychological life, as we have said, oscillates normally between these two extremes" (MM 168/307). Criticizing associationism, with a focus on the Humean theory of contiguity and resemblance between ideas, Bergson claims, "The essence of the general idea, in fact, is to be unceasingly going backwards and forwards between the plane of action and that of pure memory"; in other words, "the general idea oscillates continually" (MM 161/301) between the summit and the base of the cone.

32. Dreams are, for Bergson, caused by the "exaltation" of memory with the removal of the nervous system's inhibitory controls. When we sleep, and in some cases of damage to the nervous system, he says, "everything happens as if attention detached itself from life" (MM 174/312). See also Bergson, *The World of Dreams*.

33. It is a bit of an oversimplification to speak of a double movement, because Bergson describes both a movement of contraction and expansion, and a movement of rotation. See Lawlor, *The Challenge of Bergsonism*, 50–53.

34. Bergson symbolizes the different degrees of tension on the diagram of the cone with sections parallel to the plane of action and the base of the cone (MM 162/302). Each section represents "the whole of our past, but in a more or less contracted state." Deleuze, *Bergsonism*, 61. This distinguishes the cone of memory from Husserl's diagram of the "running-off phenomena" of consciousness, in which the horizontal line represents a chronological succession and the vertical axis represents the "sinking away" of events into the past. See Husserl, *The Phenomenology of Internal Time-Consciousness*, 49.

35. This double-movement implies not only that the past survives, but also that we are continually outside of ourselves in it, so to speak. David Couzens Hoy writes: "Heidegger's deliberate lack of engagement with Bergson on the topic of temporality is disappointing, then, because Heidegger's notion of the ecstasies of temporality seems in many ways more like Bergson's cone than Husserl's linear diagram. . . . Note that there is not one direction for all of time, but each ecstasis has its own directionality. The ecstasis can move outward toward the past by 'relaxing' the focus on the need for present action. Or it can enhance present action by focusing on the near future in a more 'contracted' way." Hoy, *The Time of Our Lives*, 197.

36. See Deleuze, *Bergsonism*, 60–61.

37. According to Deleuze, "*Matter and Memory* goes furthest in the affirmation of a radical plurality of durations: The universe is made up of modifications, disturbances, changes of tension and of energy, and nothing else. . . . Psychological duration, our duration, is now only one case among others." Deleuze, *Bergsonism*, 76.

38. In a vivid example, Bergson describes this openness and our ability to distinguish several rhythms at once: "When we are sitting on the bank of a river, the flowing of the water, the gliding of a boat or the flight of a bird, the uninterrupted murmur of our deep life, are for us three different things or a single one, at will. . . ." Henri Bergson, *Duration and Simultaneity*, trans. Leon Jacobson (Indianapolis: Bobbs-Merrill, 1965), 52; quoted in Deleuze, *Bergsonism*, 80.

39. In other words, "for Bergson, *the sense of being is neither the present nor consciousness*." Lawlor, *The Challenge of Bergsonism*, 49.

40. As Deleuze stresses, Bergson's theory of memory entails that the present "*is* not, but it acts," while the past "has ceased to act or to be useful" yet "it IS in the full sense of the word." Deleuze, *Bergsonism*, 55.

41. Heidegger makes this comparison himself in *The Metaphysical Foundations of Logic* (SS 1928), to which I return in the Conclusion.

Conclusion

1. Heidegger gives Husserl credit for seeking primordial time and discovering the intentional structures of time-consciousness, but he laments that, even for Husserl, "time gets understood as something immanent . . . something internal, 'in the subject' " (MFL 204/264).

2. Heidegger's remark about exactness evokes the opening lines of Bergson's *The Creative Mind*: "What philosophy has lacked most of all is precision. Philosophical systems are not cut to the measure of the reality in which we live; they are too wide for reality" (CM 11/1253).

3. This image of duration appears in "Introduction to Metaphysics," where Bergson writes: "It is, if you like, the unrolling of a spool, for there is no living being who does not feel himself coming little by little to the end of his span; and living consists in growing old. But it is just as much a continual winding, like that of thread into a ball, for our past follows us, becoming larger and larger with the present it picks up on its way; and consciousness means memory" (CM 164/1397).

4. See Hyppolite, "Various Aspects of Memory in Bergson," 114–15.

5. See, for example, *Creative Evolution*, 316/761ff and 345/787ff. Bergson concludes that what ancient and modern metaphysics have in common is this: "*For both, reality as well as truth are integrally given in eternity*" (354/794).

6. As Deleuze has shown, this only becomes clear if we consider the development of Bergson's thought beyond *Time and Free Will*, as "[d]uration seemed to him to be less and less reducible to a psychological experience and became instead the variable essence of things, providing the theme of a complex ontology." Deleuze, *Bergsonism*, 34. While Deleuze stresses Bergson's differences from phenomenology, he nevertheless draws attention to his proximity to Heidegger on this point: "Bergson initially discovered duration as consciousness. But further study of consciousness led him to demonstrate that it only existed in so far as it opened itself upon a whole, by coinciding with the opening up of a whole. Similarly for the living being . . . it is open upon a world, and the world, the universe, is itself the Open." Deleuze, *Cinema I*, 9–10.

7. As Ed Casey explains: "Bergson's celebrated critique of the 'spatialization' of time is just as much a critique of its 'placialization,' its representation as a set of densely juxtaposed positions on a time-line. Yet it is as important to separate place from space (construed as a homogeneous and isotropic medium) as it is to distinguish such space from true time (grasped as heterogeneous anisotropic multiplicity). In failing to make the first separation and by his insistence on the second, Bergson unwittingly falls prey to the modernist myth that place can be discounted and set aside for the sake of space or time." Edward S. Casey, *Getting Back into Place: Toward a Renewed Understanding of the Place-World* (Bloomington: Indiana University Press, 2009), 9–10.

8. According to Deleuze, beginning with *Matter and Memory*, Bergson develops a more nuanced view not only of duration but also of space, which "seemed to him to be less and less reducible to a fiction separating us from this psychological reality [of duration], rather, it was itself grounded in being." Deleuze, *Bergsonism*, 35; see also 86–89.

9. In this regard, many of Bergson's critics have overemphasized his anti-intellectualism and misrepresented his attitude toward the empirical sciences. In his defense, Bergson writes, "what I wanted was a philosophy which would submit to the control of science and which in turn could enable science to progress. And I

think it can be said that I found it, since psychology, neurology, pathology, biology have become more open to my views, once judged to be paradoxical" (CM 66/1308; see also CM 37–45/1277–89).

10. Like Heidegger, Merleau-Ponty is highly critical of Bergsonian intuition, which he interprets as "coincidence" or "fusion" with the things under consideration. See especially *The Visible and the Invisible*, 121–9. He also criticizes Bergson for failing to think about history. See, for example, *Signs*, 187–8. Interestingly, though, Merleau-Ponty confesses to a certain appropriation of Bergson's view of the historicality of truth: "What we call expression is only another formula for the phenomenon to which Bergson continually returns—the retroactive effect of the true. The experience of the true cannot keep from projecting itself back into the time which preceded it. . . . to think an idea as true, implies that we arrogate to ourselves the right of recovering the past, either to treat it as an anticipation of the present, or at least to place the pat and the present in the same world." Merleau-Ponty, *Signs*, 29.

11. Ironically, Levinas employs the concept of duration in a critical reading of Heidegger on death and time, arguing that it "avoids ideas of flux and flowing, which make us think of liquid substance and announce the possibility of a measure of time (time measured, or clock time, is not the authentic time). As *temporalization—Zeitigung*—the word 'duration' avoids all these misunderstandings and avoids the confusion between what flows *within* time and time itself." Levinas, *God, Death, and Time*, 7. The developments of Bergson's thinking in *Matter and Memory* notwithstanding, he does not abandon the image of time as flowing, that is, as continuous and irreversible. However, he usually supplements this level of description with another that emphasizes not just change or movement, but also the movement of memory, namely expansion (or the past "swelling") and contraction (the past "gnawing into" the future). See, for example, CE 1–7/495–500 and CM 162–8/1396–402.

12. Levinas regards Bergson's conception of the new, which he develops after *Matter and Memory*, as a crucial aspect of his ontology. For Bergson, to say that every moment of duration differs from the one before it means that "life recommences at each instant, receiving a new sense starting from the inimitable novelty of the present which opens upon an unforeseeable future." Levinas, *Time and the Other*, 130. If duration is "unceasing creation, the uninterrupted upsurge of novelty" (CM 18/1259), then, Levinas claims, "with Bergsonism one can think the human as the explosion of being in duration." *Time and the Other*, 132.

13. See CE 272/725ff and CM 96–99/1336–8. Interestingly, Heidegger paraphrases Bergson's view when he writes, "the nothing is the negation of the totality of beings," which he counters with the claim that "the nothing is more originary than the 'not' and negation." Heidegger, "What Is Metaphysics?," trans. David Farrell Krell, in *Pathmarks*, ed. William McNeill (Cambridge: Cambridge University Press, 1998), 85–86. While Deleuze identifies Bergson's "critique of the negative" as a key strength (*Desert Islands*, 42; see also *Bergsonism*, 43–47), Levinas interprets Bergson's "refutation of the idea of nothingness" as a failure to overcome the Western

philosophical tradition (*God, Death, and Time*, 66–70), and Merleau-Ponty characterizes Bergson's philosophy as "essentially a positivism" that nonetheless shares the faults of Sartre's "negativism" (*Nature*, 64–70; see also *The Visible and the Invisible*, ed. Claude Lefort and trans. Alphonso Lingis [Evanston: Northwestern University Press, 1968], 50ff and 193–7).

14. See Heidegger, *The Fundamental Concepts of Metaphysics* (WS 1929–30).

15. The question of Bergson's contributions to a new conception of life is currently under debate. See, for example, Frédéric Worms, "Consciousness or Life? Bergson between Phenomenology and Metaphysics," trans. Robert Vallier and John Nale, in Michael R. Kelly, ed., *Bergson and Phenomenology*, 245–57; Renaud Barbaras, "The Failure of Bergsonism," ibid., 258–72; and Alia Al-Saji, "Life as Vision: Bergson and the Future of Seeing Differently," ibid., 148–73.

Bibliography

Works by Bergson

Bergson, Henri. "Aristotle's Concept of Place." Translated by John K. Ryan. In *Ancients and Moderns: Studies in Philosophy and the Philosophy of History*, vol. 5, 20–72. Washington, DC: The Catholic University of America Press, 1970.

———. *Henri Bergson: Key Writings*. Edited by Keith Ansell Pearson and John Mullarkey. London: Continuum, 2002.

———. *Creative Evolution*. Translated by Arthur Mitchell. New York: Modern Library, 1944.

———. *The Creative Mind: An Introduction to Metaphysics*. Translated by Mabelle L. Andison. New York: Citadel Press, 1992.

———. *Duration and Simultaneity*. Translated by Leon Jacobson. Indianapolis: Bobbs-Merrill, 1965.

———. *Laughter: An Essay on the Meaning of the Comic*. Translated by Cloudesley Bereton and Fred Rothwell. London: MacMillan and Company, 1911.

———. *Matter and Memory*. Translated by N. M. Paul and W. S. Palmer. New York: Zone Books, 1991.

———. *Melanges*. Edited by André Robinet. Paris: Presses Universitaires de France, 1972.

———. *Mind-Energy*. Translated by H. Wildon Carr. New York: Henry Holt & Co., 1920.

———. *Œuvres*. Edited by André Robinet. Paris: Presses Universitaires de France, 1959.

———. *Time and Free Will: An Essay on the Immediate Data of Consciousness*. Translated by F. L. Pogson. London: George Allen & Unwin, 1910.

———. *The Two Sources of Morality and Religion*. Translated by R. Ashley Audra and Cloudesley Brereton. Notre Dame, IN: University of Notre Dame Press, 1977.

———. *The World of Dreams*. Translated by Wade Baskin. New York: Philosophical Library, 1958.

Works on Bergson

Al-Saji, Alia. "The Memory of Another Past: Bergson, Deleuze, and a New Theory of Time." *Continental Philosophy Review* 37 (2004): 203–239.

———. "'A Past Which Has Never Been Present': Bergsonian Dimensions in Merleau-Ponty's Theory of the Prepersonal." *Research in Phenomenology* 38 (2008): 41–71.

———. "A Phenomenology of Critical-Ethical Vision: Merleau-Ponty, Bergson, and the Question of Seeing Differently." *Chiasmi International* 11 (2009): 375–398.

———. "The Reality of the Virtual: Bergson and Deleuze." *MLN* 120 (2006): 1112–1127.

———. "The Temporality of Life: Merleau-Ponty, Bergson, and the Immemorial Past." *The Southern Journal of Philosophy* 45 (2007): 177–206.

Ansell Pearson, Keith. *Germinal Life: The Difference and Repetition of Deleuze.* London: Routledge, 1999.

———. *Philosophy and the Adventure of the Virtual: Bergson and the Time of Life.* London: Routledge, 2002.

Antliff, Mark. *Inventing Bergson: Cultural Politics and the Parisian Avant-Garde.* Princeton: Princeton University Press, 1993.

Ardoin, Paul, S. E. Gontarski, and Laci Mattison, eds. *Understanding Bergson, Understanding Modernism.* New York: Bloomsbury, 2013.

Barnard, G. William. *Living Consciousness: The Metaphysical Vision of Henri Bergson.* Albany: State University of New York Press, 2012.

Bernet, Rudolf. "A Present Folded Back on the Past (Bergson)." *Research in Phenomenology* 35 (2005): 55–76.

Borradori, Giovanna. "On the Presence of Bergson in Deleuze's Nietzsche. *Philosophy Today* 43 (1999): 140–145.

———. "The Temporalization of Difference: Reflections on Deleuze's Interpretation of Bergson." *Continental Philosophy Review* 34 (2001): 1–20.

Boundas, Constantin V. "Deleuze-Bergson: An Ontology of the Virtual." In *Deleuze: A Critical Reader*, edited by Paul Patton, 81–105. Oxford: Blackwell, 1996.

Burwick, Frederick, and Paul Douglass. *The Crisis in Modernism: Bergson and the Vitalist Controversy.* Cambridge: Cambridge University Press, 1992.

Crocker, Stephen. "The Oscillating Now: Heidegger on the Failure of Bergsonism." *Philosophy Today* 41 (1997): 405–423.

———. "The Past Is to Time What the Idea Is to Thought or, What Is General in the Past in General?" *Journal of the British Society for Phenomenology* 35 (2004): 42–53.

Deleuze, Gilles. *Bergsonism.* Translated by Hugh Tomlinson and Barbara Habberjam. New York: Zone Books, 1991.

———. "Bergson: 1859–1941." In *Desert Islands and Other Texts (1953–1974)*, edited by David Lapoujade and translated by Mike Taormina, 22–31. Los Angeles: Semiotext(e), 2003.

———. "Bergson's Conception of Difference." In *Desert Islands and Other Texts (1953–1974)*, edited by David Lapoujade and translated by Mike Taormina, 32–51. Los Angeles: Semiotext(e), 2003.

———. *Cinema 1: The Movement-Image*. Translated by Hugh Tomlinson and Barbara Habberjam. Minneapolis: University of Minnesota Press, 1986.

———. *Cinema 2: The Time-Image*. Translated by Hugh Tomlinson and Robert Galeta. Minneapolis: University of Minnesota Press, 1989.

Durie, Robin. "Splitting Time: Bergson's Philosophical Legacy." *Philosophy Today* 44 (2000): 152–168.

———. "Wandering among Shadows: The Discordance of Time in Levinas and Bergson." *The Southern Journal of Philosophy* 48 (2010): 371–392.

Gilson, Bernard. *La revision Bergsonienne de l'esprit*. Paris: Vrin, 1996.

Grosz, Elizabeth. "Merleau-Ponty, Bergson, and the Question of Ontology." In *Intertwinings: Interdisciplinary Encounters with Merleau-Ponty*, edited by Gail Weiss, 13–29. Albany: State University of New York Press, 2008.

———. *The Nick of Time: Politics, Evolution, and the Untimely*. Durham: Duke University Press, 2004.

———. *Time Travels: Feminism, Nature, Power*. Durham: Duke University Press, 2005.

Guerlac, Suzanne. *Thinking in Time: An Introduction to Henri Bergson*. Ithaca: Cornell University Press, 2006.

Gunter, P. A. Y., ed. *Bergson and the Evolution of Physics*. Knoxville: University of Tennessee Press, 1969.

———, ed. *Henri Bergson: A Bibliography*. Bowling Green, OH: Philosophy Documentation Center, Bowling Green State University, 1974.

———. "New Bergsons." *Continental Philosophy Review* 37 (2004): 263–270.

Hammer, Taylor. "Difference and Creativity: Virtuality and Actualization in Deleuze's Reading of Bergson." *Philosophy Today* 51 (2007): 60–68.

Hanna, Thomas, ed. *The Bergsonian Heritage*. New York: Columbia University Press, 1962.

Hyppolite, Jean. "Various Aspects of Memory in Bergson." Translated by Athena V. Colman. In *The Challenge of Bergsonism: Phenomenology, Ontology, Ethics*, edited by Leonard Lawlor, 112–127. London: Continuum, 2003.

Jankélévitch, Vladimir. *Henri Bergson*. Paris: Presses Universitaires de France, 1999.

Kelly, Michael R., ed. *Bergson and Phenomenology*. London: Palgrave MacMillan, 2010.

———. "Husserl, Deleuzean Bergsonism and the Sense of the Past in General." *Husserl Studies* 24 (2008): 15–30.

Kolakowski, Leszek. *Bergson*. South Bend, IN: St. Augustine's Press, 2001.

Lacy, A. R. *Bergson*. New York: Routledge, 1989.

Lawlor, Leonard. "Bergson Revisited." *Symposium* 10 (2006): 35–52.

———. *The Challenge of Bergsonism: Phenomenology, Ontology, Ethics*. London: Continuum, 2003.

———. "The Ontology of Memory: Bergson's Reversal of Platonism." *Epoché* 8 (2003): 69–102.

———. "What Immanence? What Transcendence? The Prioritization of Intuition over Language in Bergson." *Journal of the British Society for Phenomenology* 35 (2004): 24–41.

Lefebvre, Alexandre. *The Image of Law: Deleuze, Bergson, Spinoza*. Stanford: Stanford University Press, 2008.

Lefebvre, Alexandre, and Melanie White, eds. *Bergson, Politics, and Religion*. Durham: Duke University Press, 2012.

Marrati, Paola. "The Natural Cyborg: The Stakes of Bergson's Philosophy of Evolution." *The Southern Journal of Philosophy* 48 (2010): 3–17.

———. "Time, Life, Concepts: The Newness of Bergson." *MLN* 120 (2006): 1099–1111.

Merleau-Ponty, Maurice. "Bergson in the Making." In *Signs*, translated by Richard C. McCleary, 182–191. Evanston: Northwestern University Press, 1964.

———. "Bergson." In *In Praise of Philosophy and Other Essays*, translated by John Wild, James Edie, and John O'Neill, 9–33. Evanston: Northwestern University Press, 1968.

———. *The Incarnate Subject: Malebranche, Biran and Bergson on the Union of Body and Soul*. Edited by Andrew G. Bjelland Jr. and Patrick Burke. Translated by Paul B. Milan. Amherst, NY: Humanity Books, 2001.

Moore, F. C. T. *Bergson: Thinking Backwards*. Cambridge: Cambridge University Press, 1996.

Morris, David. "Bergsonian Intuition, Husserlian Variation, Peirceian Abduction: Toward a Relation Between Method, Sense, and Nature." *The Southern Journal of Philosophy* 43 (2005): 267–298.

Moulard-Leonard, Valentine. *Bergson-Deleuze Encounters: Transcendental Experience and the Thought of the Virtual*. Albany: State University of New York Press, 2008.

Muldoon, Mark S. *Tricks of Time: Bergson, Merleau-Ponty and Ricoeur in Search of Time, Self and Meaning*. Pittsburgh: Duquesne University Press, 2006.

Mullarkey, John. *Bergson and Philosophy*. Notre Dame, IN: University of Notre Dame Press, 1999.

———. "Forget the Virtual: Bergson, Actualism, and the Refraction of Reality." *Continental Philosophy Review* 37 (2004): 469–493.

———, ed. *The New Bergson*. Manchester: Manchester University Press, 1999.

Olkowski, Dorothea. "Merleau-Ponty and Bergson: The Character of the Phenomenal Field." In *Merleau-Ponty: Difference, Materiality, Painting*, edited by Véronique M. Fóti, 27–36. Atlantic Highlands, NJ: Humanities Press, 1996.

———. "The End of Phenomenology: Bergson's Interval in Irigaray." *Hypatia* 15 (2000): 73–91.

Riquier, Camille. *Archéologie de Bergson: Temps et métaphysique*. Paris: Presses Universitaires de France, 2009.

Robbins, Stephen B. "Bergson and the Holographic Theory of Mind." *Phenomenology and the Cognitive Sciences* 5 (2006): 365–394.

———. "Bergson, Perception, and Gibson. *Journal of Consciousness Studies* 7 (2000): 23–45.

Schatzki, Theodore R. "The Time of Activity." *Continental Philosophy Review* 39 (2006): 155–182.

Sinclair, Mark. "Inheritance, Originality, and the Will: Bergson and Heidegger on Creation." *International Journal for Philosophical Studies*, forthcoming.

———. "Is Habit the 'Fossilized Residue of a Spiritual Activity'? Ravaisson, Bergson, Merleau-Ponty." *Journal of the British Society for Phenomenology* 42 (2011): 33–52.

Smith, Steven G. "The Mind-Matter Inversions: Bergson's Conception of Mental and Material Actuality." *The Southern Journal of Philosophy* 40 (2002): 295–314.

Soulez, Phillipe. *Bergson Politique*. Paris: Presses Universitaires de France, 1989.

———, and Frédéric Worms. *Bergson: Biographie*. Paris: Presses Universitaires de France, 2002.

Toadvine, Ted. "Nature and Negation: Merleau-Ponty's Reading of Bergson." *Chiasmi International* 2 (2000): 107–118.

Trifonova, Temenuga. "Matter-Image or Image-Consciousness: Bergson contra Sartre." *Janus Head* 6 (2003): 80–114.

Winkler, Rafael. "Husserl and Bergson on Time and Consciousness." In *Analecta Husserliana* 90 (2006): 93–115.

Worms, Frédéric, ed. *Annales bergsoniennes, tome I: Bergson dans le siècle*. Paris: Presses Universitaires de France, 2002.

———, ed. *Annales bergsoniennes, tome II: Bergson, Deleuze, la phénoménologie*. Paris: Presses Universitaires de France, 2004.

———. *Introduction à* Matière et mémoire *de Bergson: suivie d'une brève introduction aux autres livres de Bergson*. Paris: Presses Universitaires de France, 1997.

———. "Is Life the Double Source of Ethics? Bergson's Ethical Philosophy between Immanence and Transcendence." *Journal of the British Society for Phenomenology* 35 (2004): 82–88.

———. *Le Vocabulaire de Bergson*. Paris: Ellipses, 2000.

Works by Heidegger

Heidegger, Martin. *Basic Problems of Phenomenology: Winter Semester 1919/1920*. Translated by Scott M. Campbell. London: Bloomsbury, 2013. (GA 58)

———. *The Basic Problems of Phenomenology*. Translated by Albert Hofstadter. Bloomington: Indiana University Press, 1982. (GA 24)

———. *Being and Time*. Translated by J. Macquarrie and E. Robinson. New York: Harper & Row, 1962. (GA 2)

———. *The Concept of Time*. Translated by William McNeill. Oxford: Blackwell, 1992. (GA 64)

———. *The Concept of Time: The First Draft of* Being and Time. Translated by Ingo Farin with Alex Skinner. London: Continuum, 2011. (GA 64)

———. *The Fundamental Concepts of Metaphysics: World, Finitude, Solitude*. Translated by William McNeill and Nicholas Walker. Bloomington: Indiana University Press, 1995. (GA 29/30)

———. *History of the Concept of Time*. Translated by Theodore Kisiel. Bloomington: Indiana University Press, 1985. (GA 20)
———. *Kant and the Problem of Metaphysics*. Translated by Richard Taft. Bloomington: Indiana University Press, 1990. (GA 3)
———. *Logic: The Question of Truth*. Translated by Thomas Sheehan. Bloomington: Indiana University Press, 2010. (GA 21)
———. *The Metaphysical Foundations of Logic*. Translated by Michael Heim. Bloomington: Indiana University Press, 1992. (GA 26)
———. *On Time and Being*. Translated by Joan Stambaugh. New York: Harper & Row, 1972.
———. *Pathmarks*. Edited by William McNeill. Cambridge: Cambridge University Press, 1998. (GA 9)
———. *Phenomenology of Intuition and Expression*. Translated by Tracy Colony. London: Continuum, 2010. (GA 59)
———. *The Phenomenology of Religious Life*. Translated by Matthias Fritsch and Jennifer Anna Gosetti-Ferencei. Bloomington: Indiana University Press, 2004. (GA 60)
———. *Supplements: From the Earliest Essays to* Being and Time *and Beyond*. Edited by John van Buren. Albany: State University of New York Press, 2002.
———. *Towards the Definition of Philosophy*. Translated by Ted Sadler. London: Continuum, 2008. (GA 56/57)

Works on Heidegger

Beistegui, Miguel de. *Heidegger and the Political: Dystopias*. London: Routledge, 1998.
———. *The New Heidegger*. London: Continuum, 2005.
———. *Truth and Genesis: Philosophy as Differential Ontology*. Bloomington: Indiana University Press, 2004.
Bernasconi, Robert. "Heidegger and the Invention of the Western Philosophical Tradition." *Journal of the British Society for Phenomenology* 26 (1995): 240–254.
———. *Heidegger in Question: The Art of Existing*. Atlantic Highlands, NJ: Humanities Press International, 1993.
———. "Heidegger's Destruction of *Phronesis*," *The Southern Journal of Philosophy* 28 (1990): 127–147.
———. *The Question of Language in Heidegger's History of Being*. Atlantic Highlands, NJ: Humanities Press International, 1989.
———. "Repetition and Tradition: Heidegger's Destructuring of the Distinction between Essence and Existence in *Basic Problems of Phenomenology*." In *Reading Heidegger from the Start*, edited by Theodore Kiesel and John Van Buren, 123–136. Albany: State University of New York Press, 1994.

Blattner, William D. *Heidegger's Temporal Idealism*. Cambridge: Cambridge University Press, 1999.
Boer, Karin de. *Thinking in the Light of Time: Heidegger's Encounter with Hegel*. Albany: State University of New York Press, 2000.
Brogan, Walter. "Double Archê: Heidegger's Reading of Aristotle's Kinetic Ontology." *Angelaki* 2 (2006): 85–93.
Campbell, Scott M. *The Early Heidegger's Philosophy of Life: Facticity, Being, and Language*. New York: Fordham University Press, 2012.
Chanter, Tina. "Heidegger's Understanding of the Aristotelian Concept of Time." In *Interrogating the Tradition: Hermeneutics and the History of Philosophy*, edited by Charles E. Scott and John Sallis, 131–157. Albany: State University of New York Press, 2000.
———. "Metaphysical Presence: Heidegger on Time and Eternity." In *Ethics and Danger: Essays on Heidegger and Continental Thought*, edited by Arleen B. Dallery and Charles E. Scott. Albany: State University of New York Press, 1992.
———. *Time, Death and the Feminine: Levinas with Heidegger*. Stanford: Stanford University Press, 2001.
Crocker, Stephen. "The Oscillating Now: Heidegger on the Failure of Bergsonism." *Philosophy Today* 41 (Fall 1997): 405–423.
Dahlstrom, Daniel O. "Heidegger's Concept of Temporality: Reflections on a Recent Criticism." *Review of Metaphysics* 49 (Summer 1995): 95–115.
Dallery, Arleen B., and Charles E. Scott, eds. *Ethics and Danger: Essays on Heidegger and Continental Thought*. Albany: State University of New York Press, 1992.
Dastur, Françoise. "The Ekstatico-horizonal Constitution of Temporality." In *Critical Heidegger*, edited by Christopher Macann, 158–170. London: Routledge, 1996.
———. *An Essay on Finitude*. Translated by John Llewelyn. London: Athlone Press, 1996.
———. *Heidegger and the Question of Time*. Translated by François Raffoul and David Pettigrew. Atlantic Highlands, NJ: Humanities Press International, 1998.
———. *Telling Time: Sketch of a Phenomenological Chrono-Logy*. Translated by Edward Bullard. London: Athlone Press, 2000.
Derrida, Jaques. *Of Spirit: Heidegger and the Question*. Translated by Geoffrey Bennington and Rachel Bowlby. Chicago: University of Chicago Press, 1989.
———. "*Ousia* and *Grammē*: Note on a Note from *Being and Time*." In *Margins of Philosophy*, translated by Alan Bass, 29–67. Chicago: University of Chicago Press, 1982.
Ellis, John. "Heidegger, Aristotle, and Time in *Basic Problems* 19." In *Interrogating the Tradition: Hermeneutics and the History of Philosophy*, edited by Charles E. Scott and John Sallis, 159–178. Albany: State University of New York Press, 2000.

Heinz, Marion. *Zeitlichkeit und Temporalität: die Konstitution der Existenz und die Grundlegung einer temporalen Ontologie im Frühwerk Martin Heideggers*. Amsterdam: Rodopi, 1982.

Kisiel, Theodore. *The Genesis of Heidegger's* Being and Time. Berkeley: University of California Press, 1993.

———, and John van Buren, eds. *Reading Heidegger from the Start: Essays in His Earliest Thought*. Albany: State University of New York Press, 1994.

———, and Thomas Sheehan, eds. *Becoming Heidegger: On the Trail of His Early Occasional Writings, 1910–1927*. Evanston: Northwestern University Press, 2007.

Krell, David Farrell. *Daimon Life: Heidegger and Life-Philosophy*. Bloomington: Indiana University Press, 1992.

Llewelyn, John. "The 'Possibility' of Heidegger's Death." *Journal of the British Society for Phenomenology* 14 (May 1983): 127–138.

Luchte, James. *Heidegger's Early Philosophy: The Phenomenology of Ecstatic Temporality*. London: Continuum, 2008.

Marx, Werner. *Heidegger and the Tradition*. Evanston: Northwestern University Press, 1971.

McMullin, Irene. "Sharing the 'Now': Heidegger and the Temporal Co-constitution of the World." *Continental Philosophy Review* 42 (2009): 201–220.

McNeil, William. *The Glance of the Eye: Heidegger, Aristotle, and the Ends of Theory*. Albany: State University of New York Press, 1999.

———. *The Time of Life: Heidegger and Ēthos*. Albany: State University of New York Press, 2006.

Nenon, Thomas. "Ethics between Tradition and a New Beginning." *Research in Phenomenology* 27 (1997): 199–206.

———. "Originary Temporality, the Origin of Modality, and the Question of the Limits of Toleration in Heidegger's *Being and Time*." In *Interpretando la Experiencia de la Tolerancia*, edited by C. Monteagudo and R. Rizo-Patron. Forthcoming.

Pöggeler, Otto. *Martin Heidegger's Path of Thinking*. Translated by Daniel Magurshak and Sigmund Barber. Atlantic Highlands, NJ: Humanities Press International, 1987.

———. *The Paths of Heidegger's Life and Thought*. Translated by John Bailiff. Amherst, NY: Humanity Books, 1998.

Protevi, John. *Time and Exteriority: Aristotle, Heidegger, Derrida*. London: Associated University Presses, 1994.

Raffoul, François. *Heidegger and the Subject*. Translated by David Pettigrew and Gregory Recco. Atlantic Highlands, NJ: Humanities Press International, 1998.

Richardson, William J. *Heidegger: Through Phenomenology to Thought*. New York: Fordham University Press, 2003.

Ricoeur, Paul. "Temporality, Historicality, Within-Time-Ness: Heidegger and the 'Ordinary' Concept of Time." In *Time and Narrative,* vol. 3, translated by

Kathleen Blamey and David Pellauer, 60–96. Chicago: University of Chicago Press, 1988.

Sallis, John. *Delimitations: Phenomenology and the End of Metaphysics*. Bloomington: Indiana University Press, 2005.

———. *Echoes: After Heidegger*. Bloomington: Indiana University Press, 1990.

———, ed. *Reading Heidegger: Commemorations*. Bloomington: Indiana University Press, 1993.

Scott, Charles E. *The Time of Memory*. Albany: State University of New York Press, 1999.

———, and John Sallis, eds. *Interrogating the Tradition: Hermeneutics and the History of Philosophy*. Albany: State University of New York Press, 2000.

Scott, David. "The 'Concept of Time' and the 'Being of the Clock': Bergson, Einstein, Heidegger, and the Interrogation of the Temporality of Modernism." *Continental Philosophy Review* 39 (2006): 83–213.

Seyppel, J. H. "A Criticism of Heidegger's Time Concept with Reference to Bergson's '*Duree*.'" *Revue Internationale de Philosophie* 10 (1956): 503–508.

Sherover, Charles. *Heidegger, Kant and Time*. Bloomington: Indiana University Press, 1971.

Taminiaux, Jacques. *Heidegger and the Project of Fundamental Ontology*. Translated by Michael Gendre. Albany: State University of New York Press, 1991.

Van Buren, John. *The Young Heidegger: Rumor of the Hidden King*. Bloomington: Indiana University Press, 1994.

Von Herrman, Friedrich-Wilhelm. "*Being and Time* and *The Basic Problems of Phenomenology*." Translated by Parvis Emad and Kenneth Maly. In *Reading Heidegger: Commemorations*, edited by John Sallis, 118–135. Bloomington: Indiana University Press, 1993.

White, Carol J. *Time and Death: Heidegger's Analysis of Finitude*. London: Ashgate, 2005.

Wood, David. "Reiterating the Temporal: Toward a Rethinking of Heidegger on Time." In *Reading Heidegger: Commemorations*, edited by John Sallis, 136–159. Bloomington: Indiana University Press, 1993.

———. *Time After Time*. Bloomington: Indiana University Press, 2007.

Other Works

Bachelard, Gaston. *The Dialectic of Duration*. Translated by Mary McAllester Jones. Manchester: Clinamen Press, 2000.

Barbaras, Renaud. *Le tournant de l'expérience: Recherches sur la philosophie de Merleau-Ponty*. Paris: J. Vrin, 1997.

Barnes, Jonathan, ed. *The Complete Works of Aristotle*. Princeton: Princeton University Press, 1984.

Brough, John B., and Lester Embree, eds. *The Many Faces of Time*. Dordrecht: Kluwer Academic Publishers, 2000.

Carr, David. *Time, Narrative, and History*. Bloomington: Indiana University Press, 1986.

Casey, Edward S. *Remembering: A Phenomenological Study*. 2nd ed. Bloomington: Indiana University Press, 2000.

———. *Getting Back into Place: Toward a Renewed Understanding of the Place-World*. Bloomington: Indiana University Press, 2009.

Deleuze, Gilles. *Difference and Repetition*. Translated by Paul Patton. New York: Columbia University Press, 1994.

———. *Logic of Sense*. Translated by Mark Lester with Charles Stivale. New York: Columbia University Press, 1990.

Derrida, Jacques. *Voice and Phenomenon: Introduction to the Problem of the Sign in Husserl's Phenomenology*. Translated by Leonard Lawlor. Evanston: Northwestern University Press, 2011.

Dilthey, Wilhelm. *Selected Works Volume I: Introduction to the Human Sciences*. Edited by Rudolf A. Makkreel and Frithjof Rodi. Princeton: Princeton University Press, 1991.

———. *Selected Works Volume III: The Formation of the Historical World in the Human Sciences*. Edited by Rudolf A. Makkreel and Frithjof Rodi. Princeton: Princeton University Press, 2002.

Durie, Robin, ed. *Time and the Instant*. Manchester: Clinamen, 2000.

Gallagher, Shaun. *The Inordinance of Time*. Evanston: Northwestern University Press, 1998.

Giroux, Laurent. *Durée Pure et Temporalité: Bergson et Heidegger*. Montreal: Ballarmin, 1971.

Grosz, Elizabeth. *Becomings: Explorations in Time, Memory, and Futures*. Ithaca: Cornell University Press, 1999.

Gutting, Gary. *French Philosophy in the Twentieth Century*. Cambridge: Cambridge University Press, 2001.

Hegel, Georg Wilhelm Friedrich. *The Jena System of 1804–1805: Logic and Metaphysics*. Translated by J. Burbidge and George di Giovanni. Kingston: McGill-Queen's Press, 1986.

———. *Hegel's Philosophy of Nature: Being Part Two of the Encyclopaedia of the Philosophical Sciences (1830)*. Translated by A. V. Miller. Oxford: Clarendon Press, 1970.

Hoy, David Couzens. *The Time of Our Lives: A Critical History of Temporality*. Cambridge, MA: The MIT Press, 2009.

Husserl, Edmund. *Ideas: General Introduction to Pure Phenomenology*. Translated by W. R. Boyce Gibson. New York: Collier Books, 1962.

———. *On the Phenomenology of the Consciousness of Internal Time (1893–1917)*. Edited by Rudolf Boehm. Translated by John B. Brough. Dordrecht: Kluwer Academic Publishers, 1991.

———. *The Phenomenology of Internal Time-Consciousness*. Edited by Martin Heidegger. Translated by James S. Churchill. Bloomington: Indiana University Press, 1928.

James, William. *The Principles of Psychology*, vol. 1. New York: Henry Holt & Co., 1890.

Kant, Immanuel. *Critique of Pure Reason*. Translated by Norman Kemp Smith. New York: St. Martin's Press, 1965.

Kierkegaard, Søren. *Fear and Trembling/Repetition*. Edited and translated by Howard V. Hong and Edna H. Hong. Princeton: Princeton University Press, 1983.

Lawlor, Leonard. *Derrida and Husserl: The Basic Problem of Phenomenology*. Bloomington: Indiana University Press, 2002.

———. *Early Twentieth-Century Continental Philosophy*. Bloomington: Indiana University Press, 2012.

———. "Temporality and Spatiality: A Note to a Footnote in Jaques Derrida's *Writing and Difference*." *Research in Phenomenology* 12 (1982): 149–166.

———. *Thinking through French Philosophy: The Being of the Question*. Bloomington: Indiana University Press, 2003.

Levinas, Emmanuel. *God, Death, and Time*. Translated by Bettina Bergo. Stanford: Stanford University Press, 2000.

———. *Time and the Other*. Translated by A. Cohen. Pittsburgh: Duquesne University Press, 1987.

Lloyd, Genevieve. *Being in Time: Selves and Narrators in Philosophy and Literature*. London: Routledge, 1993.

Marrati, Paola. *Gilles Deleuze: Cinema and Philosophy*. Translated by Alisa Hartz. Baltimore: Johns Hopkins University Press, 2008.

Merleau-Ponty, Maurice. *Nature: Course Notes from the Collège de France*. Edited by Dominique Séglard. Translated by Robert Vallier. Evanston: Northwestern University Press, 2003.

———. *Phenomenology of Perception*. Translated by Colin Smith. London: Routledge, 1962.

———. *The Visible and the Invisible*. Translated by Alphonso Lingis. Evanston: Northwestern University Press, 1968.

Sartre, Jean-Paul. *Being and Nothingness*. Translated by Hazel Barnes. New York: Washington Square Press, 1956.

———. *The Imagination*. Translated by Kenneth Williford and David Rudrauf. New York: Routledge, 2012.

Schrift, Alan D. *Twentieth-Century French Philosophy: Key Themes and Thinkers*. Malden, MA: Blackwell: 2006.

Sherover, Charles. *The Human Experience of Time: The Development of Its Philosophic Meaning*. Evanston: Northwestern University Press, 1975.

Spiegelberg, Herbert. *The Phenomenological Movement: A Historical Introduction*, 2 vols. The Hague: Martinus Nijhoff, 1971.

Turetzky, Philip. *Time*. London: Routledge, 1998.

Wood, David. *The Deconstruction of Time*. Atlantic Highlands, NJ: Humanities Press International, 1989.

Zahavi, Dan, ed. *Self-Awareness, Temporality, and Alterity: Central Topics in Phenomenology*. Dordrecht: Kluwer Academic Publishers, 1998.

Index